T0199257

# GRAPH SAMPLING

# GRAPH SAMPLING

Li-Chun Zhang

## CRC Press
Taylor & Francis Group
Boca Raton  London  New York

CRC Press is an imprint of the
Taylor & Francis Group, an **informa** business

First edition published 2022
by CRC Press
6000 Broken Sound Parkway NW, Suite 300, Boca Raton, FL 33487-2742

and by CRC Press
2 Park Square, Milton Park, Abingdon, Oxon, OX14 4RN

CRC Press is an imprint of Taylor & Francis Group, LLC

ISBN: 9781032067087 (hbk)
ISBN: 9781032067094 (pbk)
ISBN: 9781003203490 (ebk)

DOI: 10.1201/9781003203490

Publisher's note: This book has been prepared from camera-ready copy provided by the authors

*To my parents*

# Contents

# Preface

Finite population sampling has found numerous social, economic, medical, environmental and other scientific applications in the past century. The validity of inference of real populations derives from the known probability sampling design under which the sample is selected, "irrespectively of the unknown properties of the target population studied" (Neyman, 1934).

Representing a population by a graph allows one to incorporate the connections or links among the population units in addition. The links may provide effectively access to the part of the population that is the primary target, which is the case for many so-called unconventional sampling methods, such as indirect, network, line-intercept or adaptive cluster sampling. Or, one may be interested in the structure of the connections in terms of the corresponding graph properties or parameters, such as when various breadth- or depth-first non-exhaustive search algorithms are applied to obtain compressed views of large, often dynamic graphs.

Graph sampling provides a statistical approach to study real graphs from either of these perspectives. It is based on exploring the variation over all possible sample graphs (or subgraphs), which can be taken from the given population graph by means of the relevant known sampling probabilities.

The book can either be read as a research monograph or used as the basis of an advanced course for post-graduate students in statistics, mathematics and data science. It draws heavily on the development in the last 4-5 years, much of which was carried out jointly with Martina Patone and Melike Oguz-Alper.

Some of the work was done during the weeks when I visited Giovanna Ranalli at the University of Perugia, which benefitted from their Visiting Scientist Program. Finally, I thank CRC Press Focus, in particular Acquiring Editor, Rob Calver and Editorial

Assistant, Vaishali Singh, for their support in the making of this book.

**Li-Chun Zhang**
*Kongsvinger, Norway*
*August 2021*

# Author Bio

**Li-Chun Zhang** is Professor of Social Statistics at the University of Southampton, Senior Researcher at Statistics Norway, and Professor of Official Statistics at the University of Oslo. He has researched and published on topics such as finite population sampling design and coordination, graph sampling, machine learning, sample survey estimation, non-response, measurement errors, small area estimation, index number calculations, editing and imputation, register-based statistics, population size estimation, statistical matching, record linkage.

# Abbreviations

| | |
|---|---|
| HT | Horvitz-Thompson |
| OP | observation procedure |
| SRS | simple random sampling |
| BIGS | bipartite incidence graph sampling |
| IWE | incidence weighting estimator |
| HTE | Horvitz-Thompson estimator, HT-estimator |
| HH | Hansen-Hurwitz |
| RB | Rao-Blackwell |
| PIDA | probability and inverse-degree adjusted |
| RE | relative efficiency |
| LIS | line-intercept sampling |
| IID | independent and identically distributed |
| ACS | adaptive cluster sampling |
| SBS | snowball sampling |
| $T$SBS | $T$-wave snowball sampling |
| MH | Metropolis-Hastings |
| TWS | targeted walk sampling |
| TRW | targeted random walk |
| TRWS | targeted random walk sampling |
| $T$TRWS | $T$-step targeted random walk sampling |
| S3P | stationary successive sampling probability |
| AS3 | actual sampling sequence of states |
| ES3 | equivalent sampling sequence of states |

# Symbols

## SYMBOL DESCRIPTION

| | | | |
|---|---|---|---|
| $G$ | population graph $G = (U, A)$ | $A_s$ | sample of edges in $G_s$ |
| | | $\Omega_s$ | sample of motifs, study units |
| $U$ | nodes in $G$, population units | | |
| | | $s_{\mathrm{ref}}$ | reference set, parts of adjacency matrix |
| $A$ | edges in $G$ | | |
| $\Omega$ | motifs in $G$, study units in $\mathcal{B}$ | $s_0$ | initial node sample |
| | | $s$ | seed sample |
| $M$ | set of nodes | $\mathcal{B}$ | population bipartite graph $\mathcal{B} = (F, \Omega; H)$ |
| $i, j, h$ | node, sampling unit | | |
| $\kappa, \ell$ | motif, study unit | $F$ | sampling units |
| $\theta$ | graph total | $H$ | edges in $\mathcal{B}$ |
| $\mu$ | graph parameter | $\mathcal{B}_s$ | sample graph from $\mathcal{B}$, $\mathcal{B}_s = (s_0, \Omega_s; H_s)$ |
| $G_s$ | sample graph $G_s = (U_s, A_s)$ | | |
| | | $H_s$ | sample of edges in $\mathcal{B}_s$ |
| $U_s$ | sample of nodes in $G_s$ | | |

# CHAPTER 1

# General introduction

## 1.1 SAMPLING FROM FINITE POPULATIONS

Denote by $U = \{1, ..., N\}$ a *population* of size $N$. Denote by $s$ a *sample* from $U$, $s \subset U$, according to some specified method of sampling. For any $i \in U$, let $\pi_i = \Pr(i \in s)$ be the *sample inclusion probability*, where $\pi_i > 0$, which is either known before the sample is drawn or can be calculated afterwards.

Let $y_i$ be an unknown constant associated with each population unit $i$, $i \in U$. Let $Y = \sum_{i \in U} y_i$ be its population total. The Horvitz-Thompson (HT) estimator of $Y$ is given by

$$\hat{Y}_{HT} = \sum_{i \in s} \frac{y_i}{\pi_i} = \sum_{i \in U} \delta_i \frac{y_i}{\pi_i}$$

where $\delta_i = \mathbb{I}(i \in s)$. It is unbiased over hypothetically repeated sampling, denoted by $E(\hat{Y}_{HT}) = Y$. The sampling variance of $\hat{Y}_{HT}$ is given by

$$V(\hat{Y}_{HT}) = \sum_{i \in U} \sum_{j \in U} \left(\frac{\pi_{ij}}{\pi_i \pi_j} - 1\right) y_i y_j := \sum_{i \in U} \sum_{j \in U} v_{ij}$$

where $\pi_{ij} = E(\delta_i \delta_j) = \Pr(i \in s, j \in s)$ is the *joint* sample inclusion probability of $(i, j)$. Consider $V(\hat{Y}_{HT})$ as the total of $v_{ij}$ over $U \times U$, whose element $(ij)$ has inclusion probability $\pi_{ij}$ in $s \times s$. If $\pi_{ij} > 0$ for any $(ij) \in U \times U$, then an unbiased estimator of $V(\hat{Y}_{HT})$ can be given by

$$\hat{V}(\hat{Y}_{HT}) = \sum_{(ij) \in s \times s} \frac{v_{ij}}{\pi_{ij}} = \sum_{i \in s} \sum_{j \in s} \left(\frac{1}{\pi_i \pi_j} - \frac{1}{\pi_{ij}}\right) y_i y_j$$

DOI: 10.1201/9781003203490-1

## 1.2 GRAPH, MOTIF, GRAPH PARAMETER

Representing a population $U$ by a graph allows one to incorporate the connections or links among the population units in addition.

### 1.2.1 Graph

A graph $G = (U, A)$ consists of a set of nodes $U$ and a set of edges $A$, where $|U| = N$ and $|A| = R$ are the *order* and *size* of $G$, respectively.

Attaching values to $U$ or $A$ yields a *valued graph*, and $G$ is then the *structure* of the valued graph.

Let $A_{ij}$ be the set of edges from $i$ to $j$, such that $A = \bigcup_{i,j \in U} A_{ij}$. Let $a_{ij} = |A_{ij}|$. The graph is a *multigraph* if $a_{ij} > 1$ for some $i, j \in U$; it is a *simple* graph otherwise.

By default, a graph is *directed*, or a *digraph*, in this book. The out-edges of node $i$ are $A_{i+} = \bigcup_{j \in U} A_{ij}$ and the in-edges of it are $A_{+i} = \bigcup_{j \in U} A_{ji}$. The out-degree is $a_{i+} = |A_{i+}| = \sum_{j \in U} a_{ij}$ and the in-degree is $a_{+i} = |A_{+i}| = \sum_{j \in U} a_{ji}$. A graph is *undirected* if there is no distinction between in- and out-edges, $A_{ij} \equiv A_{ji}$, where $d_i = a_{i+} = a_{+i}$ is the degree of node $i$.

Two nodes $i$ and $j$ are *adjacent* if $a_{ij} + a_{ji} > 1$, in which case there exists at least one edge between them. For any edge in $A_{ij}$, $i$ is its initial node and $j$ its terminal node. An edge is *incident* to its initial and terminal nodes, and vice versa. Note that adjacency refers to relationship between nodes, as objects of the same kind, whereas incidence refers to relationship between nodes and edges, as objects of different kinds.

Let $\alpha_i$ be the *successors* of $i$, which are the terminal nodes of the out-edges from $i$; let $\beta_i$ be the *predecessors* of $i$, which are the initial nodes of the in-edges to $i$. We have $a_{i+} = |\alpha_i|$ and $a_{+i} = |\beta_i|$ for simple graphs. We have $\alpha_i \equiv \beta_i$ for undirected graphs, where $\nu_i = \alpha_i = \beta_i$ can be referred to as the *neighbourhood* of node $i$.

Where a distinction is relevant, $(ij)$ denotes an element of $U \times U$, whereas $(i, j)$ denotes a pair of nodes in $U$. Note that $(ij)$ is distinct to $(ji)$ in digraphs, whereas they are the same in undirected graphs. One can write $A_{ij} = \{(ij)\}$ if $a_{ij} = 1$, or $A_{ij} = \{(ij)_1, ..., (ij)_{a_{ij}}\}$ if $a_{ij} > 1$.

An edge incident to the same node $i$ at both ends is called a *loop*, which can sometimes be a useful representation. Whether or not loops are included in the terms and notations defined above will be a matter of convention.

## 1.2.2 Motif

Let $G(M) = \big(M, A \cap (M \times M)\big)$ be the subgraph *induced* by $M$, for $M \subseteq U$. For example, in Figure 1.1, the subgraph induced by $M = \{i_1, i_2, i_3, i_4\}$ has edges $\{(i_1 i_2), (i_2 i_3), (i_3 i_1), (i_3 i_4)\}$.

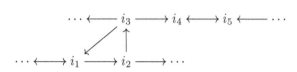

**Figure 1.1**   Illustration of subgraph, motif

The specific characteristics of $G(M)$ will be called *motif* here, denoted by $[M]$. The *order* of $[M]$ is $|M|$. For instance, $[i : a_{i+} = 3]$ defines a 1st-order motif, which is a node with out-degree 3, such as the node $i_3$ in Figure 1.1. Whereas $[i, j : a_{ij} a_{ji} = 1]$ defines a 2nd-order motif of a node pair with "mutual simple relationship", such as $G(\{i_4, i_5\})$ in Figure 1.1.

A motif $[M]$ is said to be *induced* if it can be determined based on the induced subgraph $G(M)$ itself. Thus, $[i, j : a_{ij} a_{ji} = 1]$ is an induced motif, but not $[i : a_{i+} = 3]$ which depends on $A \cap (\{i\} \times U)$ outside $G(\{i\})$.

Some examples of low-order induced motifs in undirected graphs are given in Figure 1.2.

A *component* of an undirected graph is an induced subgraph in which any two nodes are connected to each other, via a sequence of adjacent nodes, and which is connected to no other nodes in the rest of the graph. Component is not an induced motif.

As another example, a shortest path over an ordered node set $M$ is not an induced motif generally. Take ordered $M = \{i_1, i_2, i_3, i_4\}$ in Figure 1.1, whether $\{(i_1 i_2), (i_2 i_3), (i_3 i_4)\}$ is a shortest path from $i_1$ to $i_4$ depends also on $A \setminus (M \times M)$. Or, if $M = \{i_1, i_2, i_3\}$, then $\{(i_1 i_2), (i_2 i_3)\}$ is a shortest path iff $(i_1 i_3) \notin A$, based on $G(M)$.

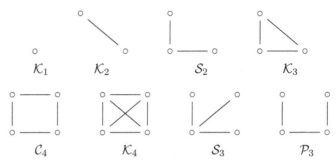

**Figure 1.2** Node ($\mathcal{K}_1$), 2-clique ($\mathcal{K}_2$), 2-star ($\mathcal{S}_2$), triangle ($\mathcal{K}_3$), 4-cycle ($\mathcal{C}_4$), 4-clique ($\mathcal{K}_4$), 3-star ($\mathcal{S}_3$), 3-path ($\mathcal{P}_3$)

### 1.2.3 Graph parameter

Let $y(M)$ be a function of $G(M)$, including the values associated with $G(M)$ in valued graphs. Let $\Omega$ contain all the relevant node sets $M$. The *graph total* of $y(M)$ over $\Omega$ is given by

$$\theta = \sum_{M \in \Omega} y(M) \tag{1.1}$$

It is said to be of the *q-th order*, if $|M| \equiv q$ for any $M \in \Omega$.

We shall refer to an arbitrary function of $\{y(M) : M \in \Omega\}$ as a *graph parameter*, denoted by $\mu$. A graph parameter has the order $q$ if $|M| \equiv q$, $\forall M \in \Omega$.

It is often convenient to let $\Omega$ be the reference set of motifs directly, which do not all need to be of the same kind, such that each $\kappa \in \Omega$ is an *occurrence* of a relevant motif in $G$. A graph total over $\Omega$ can then be given as

$$\theta = \sum_{\kappa \in \Omega} y_\kappa \tag{1.2}$$

For example, let each $\kappa$ in $\Omega$ be a distinct triangle, like the one corresponding to $G(M)$ with $M = \{i_1, i_2, i_3\}$ in Figure 1.1.

### 1.2.4 Examples

Each $M = \{i\}$ refers to a node $i \in U$. Each $M = \{i, j\}$ of two distinct nodes is called a *dyad*. Each $M = \{i, j, h\}$ of three distinct nodes is called a *triad*.

$N = order\ of\ G$  Let $\Omega = U$, and $y(i) \equiv 1$ for $i \in U$. Then $\theta = |U| = N$.

$N_d = number\ of\ degree\text{-}d\ nodes$  Let $\Omega = U$ and $y(i) = \mathbb{I}(d_i = d)$ for $i \in \Omega$ in undirected graphs.

$R = size\ of\ G$  Let $\Omega$ contain all the dyads. Let $y(i,j) = a_{ij}$ for $i, j \in U$ in digraphs, or $a_{ij}/2$ in undirected graphs. Then $\theta = R$. For any undirected graph, we have

$$2R = \sum_{i \in U} d_i = \sum_{d=1}^{D} dN_d \qquad \text{where} \quad D = \max_{i \in U} d_i$$

The 2nd-order graph total $R$ can be given as a graph parameter in terms of 1st-order totals $N_d$.

*Adjacency*  Let $y(i,j) = \mathbb{I}(a_{ij} + a_{ji} > 0)$ indicate whether $i$ and $j$ are adjacent in a simple graph. The graph parameter $\mu = R/\binom{N}{2}$ is a measure of *immediacy*. A graph of minimum immediacy consists of only isolated nodes; a graph of maximum immediacy is a *clique*, where $(ij) \in A$ for any $i, j \in U$.

*Mutual relationship*  Let $y(i,j) = \mathbb{I}(a_{ij}a_{ji} > 0)$ indicate whether $i$ and $j$ have reciprocal edges between them, in which case their relationship is *mutual*.

*Undirected triads*  For undirected simple graphs, Frank (1981) shows that there exists an explicit relationship between the mean and variance of the degree distribution and the triads of the graph. The numbers of triads of size 3, 2 and 1 are, respectively,

$$\theta_{3,3} = \sum_{i<j<h} a_{ij}a_{jh}a_{ih}$$

$$\theta_{3,2} = \sum_{i<j<h} a_{ij}a_{ih}(1 - a_{jh}) + a_{ij}a_{jh}(1 - a_{ih}) + a_{ih}a_{jh}(1 - a_{ij})$$

$$\theta_{3,1} = \sum_{i<j<h} \big(a_{ij}(1 - a_{jh})(1 - a_{ih}) + a_{ih}(1 - a_{ij})(1 - a_{jh})$$

$$+ a_{jh}(1 - a_{ij})(1 - a_{ih})\big)$$

Let

$$\mu = \sum_{d=1}^{N} dN_d/N = 2R/N$$

$$\sigma^2 = Q/N - \mu^2 \quad \text{and} \quad Q = \sum_{d=1}^{N} d^2 N_d$$

We have

$$R = \frac{1}{N-2}\left(\theta_{3,1} + 2\theta_{3,2} + 3\theta_{3,3}\right)$$

$$Q = \frac{2}{N-1}\left(\theta_{3,1} + N\theta_{3,2} + 3(N-1)\theta_{3,3}\right)$$

*Transitivity*   For an undirected simple graph, $\theta_{3,3}$ above is the total number of triangles. Let $(ij) \in A$ if a certain relationship exists between $i$ and $j$. The relationship is transitive if any three connected nodes form a triangle. One can measure the proximity to transitivity by the graph parameter $\mu = \theta_{3,3}/(\theta_{3,3} + \theta_{3,2})$.

*Directed triangles*   Let $z(j, i, h) = a_{ji}a_{ih}a_{hj}$ for ordered $(j, i, h)$ be the count of *strongly connected* triangles from $j$ via $i$ and $h$ back to $j$ in digraphs. Let $\tilde{M}$ contain all the possible orderings of a triad $M$: $(i, j, h)$, $(i, h, j)$, $(j, i, h)$, $(j, h, i)$, $(h, i, j)$ and $(h, j, i)$. The number of strongly connected triangles is given by (1.1) over all the triads, where

$$y(M) = \sum_{(i,j,h)\in\tilde{M}} z(i, j, h)/3$$

*Order-q component*   Let $\Omega_q = \{M \subseteq U : |M| = q\}$. Let $y(M) = 1$ if $G(M)$ is a component and $y(M) = 0$ otherwise. The corresponding $\theta_q$ by (1.1) is the number of components of order $q$. The number of components (of unspecified order) is the graph total given by $\theta = \sum_{q=1}^{N} \theta_q$ over $\Omega = \{M : M \subseteq U\}$. At one end, where $A = \emptyset$, there are no edges at all in the graph, we have $\theta = N = \theta_1$ and $\theta_q = 0$ for $q > 1$. At the other end, where there exists a path between any two nodes, we have $\theta = \theta_N = 1$ and $\theta_q = 0$ for $q < N$.

*Trees in a forest*   In an undirected simple graph, $[M]$ is a *tree* if the number of edges in $G(M)$ is $|M| - 1$. The graph is a *forest* if every component of it is a tree. The total number of trees in a forest-graph is the graph parameter $\mu = N - R$.

*Clique*  In an undirected simple graph, $[M]$ is a *clique* if $G(M)$ is a component, in which any two nodes are adjacent. Let $\theta$ be the graph total of cliques. As an immediacy measure, $(N - \theta)/(N - 1)$ varies between 0 and 1.

*Geodesic*  Let $G$ be connected. Let $\{i_0i_1, i_1i_2, ..., i_{q-1}i_q\}$ be a path of length $q$, over the ordered set $M = \{i_0, i_1, ..., i_q\}$. It is a *geodesic* if it is a shortest path from $i_0$ to $i_q$. Let $\Omega$ contain all such $M$, where only one of them is retained if there are more than one geodesic from $i_0$ to $i_q$. Let $y(M) = |M| - 1$. As a closeness centrality measure, let $\mu$ be the harmonic mean of $y(M)$ over $\Omega$, whose minimum value is 1 if $G$ is a clique.

## 1.3  OBSERVATION PROCEDURE

Compared to how samples are selected from finite populations, making use of the edges is a defining feature of selecting subgraphs from real graphs. The way in which the edges are used to drive subgraph selection is called the *observation procedure (OP)*.

Given an *initial sample (of nodes)* $s_0 \subset U$, the incident edges that are immediately available for any OP are either in

$$\alpha(s_0) = \bigcup_{i \in s_0} \alpha_i \quad \text{or} \quad \beta(s_0) = \bigcup_{i \in s_0} \beta_i$$

where $\alpha(s_0) = \beta(s_0) = \nu(s_0)$ for undirected graphs.

The initial sample $s_0$ can as well be given as the nodes that are incident to a subset of edges from $A$. Initial sampling of edges may be useful if the graph is known but too large to be counted or if the graph is more readily accessible via the edges.

### 1.3.1  Basic OPs

An OP is *induced* when $A_{ij}$ is observed iff both $i \in s_0$ and $j \in s_0$, or *incident-reciprocal* when $A_{ij}$ and $A_{ji}$ are both observed if either $i \in s_0$ or $j \in s_0$. Moreover, for digraphs, an incident non-reciprocal

OP is *forward* when $A_{ij}$ is observed if $i \in s_0$, or *backward* when $A_{ij}$ is observed if $j \in s_0$.

It is convenient to specify the observed edges as

$$A_s = A \cap s_{\text{ref}}$$

via a *reference set* $s_{\text{ref}}$, which explicates the parts of the *adjacency matrix* (of $a_{ij}$), which are observed given $s_0$ and the OP.

Take the graph $G$ in Figure 1.3. Let the rows and columns of the adjacency matrix be in the order $i, b, c, d$. Given $s_0 = \{b, d\}$, the reference set $s_{\text{ref}}$ is marked by $\boxed{\cdot}$ for each basic OP. We observe none of the edges in $G$ if the OP is induced, the edge $(bc)$ if it is incident forward, or $(cd)$ if incident backward, or both $(bc)$ and $(cd)$ if incident reciprocal.

Note that the observation of the motif $[M]$ requires not only observing the nodes $M$, but also it is possible to identify whether $[M]$ is the particular motif of interest. In particular, any induced motif $[M]$ is observed iff

$$M \times M \subseteq s_{\text{ref}}$$

Take e.g. the motif 2-star. To identify whether $[\{i, j, h\}]$ is a 2-star, $s_{\text{ref}}$ needs to contain $(ij)$, $(ih)$ and $(jh)$. An example where this is not the case is $i \in s_0$ and $j, h \in \alpha_i$ by incident OP, but $j, h \notin s_0$, so that the observed part of the triad is a star, but it is unclear if $a_{jh} = 0$ in $G$ because $(jh) \notin s_{\text{ref}}$.

Let $\pi_i$ and $\pi_{ij}$ be the inclusion probabilities in the initial $s_0$. Let $\bar{\pi}_i = \Pr(i \notin s_0)$ be the *exclusion probability* of $i \in U$, and let $\bar{\pi}_{ij} = \Pr(i \notin s_0, j \notin s_0)$ be that of $i, j \in U$. Similarly for $\bar{\pi}_M$. Let $\pi_{(ij)}$ be the probability of $(ij) \in A_s$ after applying the OP to $s_0$, and $\pi_{(ij)(hl)}$ that of the two edges $(ij)$ and $(hl)$ jointly.

- Induced, $s_{\text{ref}} = s_0 \times s_0$. Both $(ij) \in s_{\text{ref}}$ and $(ji) \in s_{\text{ref}}$ iff $i \in s_0$ and $j \in s_0$. Thus, $\pi_{(ij)} = \pi_{ij}$ and $\pi_{(ij)(hl)} = \pi_{ijhl}$.

- Incident-forward, $s_{\text{ref}} = s_0 \times U$. We have $(ij) \in s_{\text{ref}}$ iff $i \in s_0$. Thus, $\pi_{(ij)} = \pi_i$ and $\pi_{(ij)(hl)} = \pi_{ih}$.

- Incident-reciprocal, $s_{\text{ref}} = s_0 \times U \cup U \times s_0$. We have $(ij) \notin s_{\text{ref}}$ iff $i \notin s_0$ and $j \notin s_0$. Thus, we have $\pi_{(ij)} = 1 - \bar{\pi}_{ij}$ and we have $\pi_{(ij)(hl)} = 1 - \bar{\pi}_{ij} - \bar{\pi}_{hl} + \bar{\pi}_{ijhl}$.

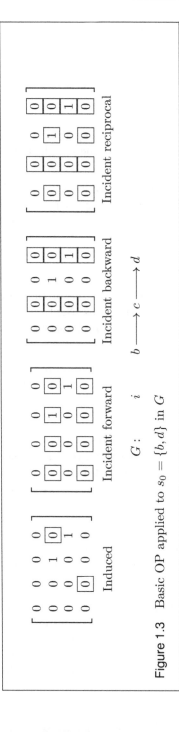

Figure 1.3  Basic OP applied to $s_0 = \{b, d\}$ in $G$

From now on, let incident mean incident-forward in case of digraphs unless otherwise specified. For undirected graphs, incident is the same as incident-forward, backward or reciprocal.

### 1.3.2 Multiplicity

There exists generally a *multiplicity (of access)* to any given motif in a graph. The multiplicity can be modified by both the OP and the sampling design of $s_0$.

For an epidemiological study in the population $U$, let a *case* be a person who would be positive when subjected to a diagnostic test. For any $i \in U$, let $y_i = 0$ if $i$ is not a case, let $y_i = 1$ if $i$ is a hospitalised case, and let $y_i = 2$ if $i$ is a non-hospitalised case. Let $\mu = \sum_{i \in U} \mathbb{I}(y_i > 0)/N$ be the parameter of interest, where $N = |U|$. Consider three designs below.

*Study-I* Let $s_0$ be an initial sample of cases who are receiving treatment at a sample of hospitals. Next, all the individuals that have been in contact with anyone in $s_0$ during the last fortnight are tracked down and tested.

*Study-II* Let $s_0$ be a simple random sample from $U$. Let $s_0'$ be the subsample of cases, where $s_0' \subseteq s_0$. All the individuals that have been in contact with anyone in $s_0'$ during the last fortnight are tracked down and tested.

*Study-III* Randomly select and test one person $i_0 \in U$. Next, select-track-test randomly one among all the individuals that have been in contact with $i_0$ during the last fortnight, denoted by $i_1$. Repeat for $i_{t+1}$ given $i_t$, where $t \geq 1$, till a fixed number of steps $n$ has been reached.

Let $G = (U, A)$ be undirected, where $a_{ij} = 1$ if two persons $i, j \in U$ had contact in the last fortnight, and $a_{ij} = 0$ otherwise. For each $i \in U$, let $(d_{0i}, d_{1i}, d_{2i})$ be the numbers of its adjacent nodes with $y = 0, 1, 2$, respectively, where $d_i = d_{0i} + d_{1i} + d_{2i}$.

In Study-I, the initial sample $s_0$ is a single-stage cluster sample from $U_1 = \{i \in U : y_i = 1\}$, and one-step tracing from each node in $s_0$ amounts to the incident OP. A node $j \in U_2 = \{i \in U : y_i = 2\}$ is selected if $d_{1j} > 1$ and $\beta_j \cap s_0 \neq \emptyset$. Whenever $|\beta_j \cap s_0| < d_{1j}$, there are nodes outside of $s_0$, which could lead to $j$ over repeated sampling. The multiplicity is $d_{1j}$ for $j \in U_2$; it is $1 + d_{1j}$ for $j \in U_1$.

In Study-II, the incident OP is said to be *adaptive*, because tracing from $i \in s_0$ is implemented only if $y_i > 0$, such that it depends on the values associated with the graph. The multiplicity of any $j \in U_1 \cup U_2$ is $1 + d_{1j} + d_{2j}$.

In Study-III, we have $s_0 = \{i_0\}$. The incident OP involves now *subsampling* of a single edge among all those incident to $i_0$. The successive *states* $\{i_0, i_1, ..., i_n\}$ form a Markov chain. The multiplicity of any $j \in U_1 \cup U_2$ is $|F_j|$, where $F_j = \{ i \in U : \psi_{ij} \leq n \}$ and $\psi_{ij}$ is the geodesic distance between $i$ and $j$.

### 1.3.3 Multiwave

We refer to *network* as a set of connected nodes, which satisfy certain specified conditions. Any node in the graph is then called a *(network) edge node* if it does not belong to the network but is adjacent to at least one node in the network; together, the edge nodes separate the network from the rest nodes. If the network condition is void, then a network would simply be a component of the graph, which has no edge nodes.

Figure 1.4 illustrates an epidemiological study, where each network is a group of cases connected by contacts, and an edge node is a noncase that has contact with at least one case.

**Figure 1.4** Illustration of network nodes (★), edge nodes (○) and others (∘)

One can repeat the incident OP *wave by wave*. The repeated OP is said to be *network exhaustive* if any network that intersects the initial sample $s_0$ will eventually be observed in its entirety.

For Study-I or II above, repeating the OP will exhaust any case network that intersects $s_0$. Since $s_0$ is only selected from the hospitalised cases in Study-I, a network has no chance of being observed if it does not contain any hospitalised cases, whereas every case network has a chance in Study-II. In Study-III, all the case networks will be exhausted, as $n \to \infty$, as long as the graph is connected and fixed.

The $T$-*wave* incident OP starting from $s_0$ is given as follows. For $t = 1, ..., T$, let

$$s_t = \alpha(s_{t-1}) \setminus \bigcup_{r=0}^{t-1} s_r$$

be the $t$-th wave sample of nodes. Let

$$s = \bigcup_{t=0}^{T-1} s_t$$

be the *seed sample*, to which the OP is applied. We would have $s_{\text{ref}} = s \times U$ for digraphs and $s_{\text{ref}} = s \times U \cup U \times s$ for undirected graphs. The OP is terminated if $s_t = \emptyset$ for some $t < T$, in which case $s_{t+1} = \cdots = s_T = \emptyset$ as well.

We have $T = 1$ in Study-I and II above, and $T = n$ in Study-III.

## 1.3.4 Multigraph

In a multigraph, incident observation from $i$ to $j$ includes all the edges in $A_{ij}$. It can often be treated as incident OP in a valued simple graph, with $a_{ij}$ from the original multigraph as the value attached to the edge $(ij)$ in the simple graph, if $a_{ij} > 0$. Below are some examples.

In an undirected multigraph $G$, let each node correspond to a parent and each edge $(ij)$ a child of $i$ and $j$. Let $A_{ii}$ represent the children of a single-parent $i$. One may be interested in the structure of such parental networks.

In a directed multigraph $G$, let each node be a locality and each edge $(ij)$ a telephone call from $i$ to $j$. Let $A_{ii}$ be the calls within the locality. One may be interested in the spatial connectivity of telephone calls.

In a directed multigraph $G$, let each node be a Twitter account and each edge $(ij)$ a tweet from $j$, which is shared again by $i$. One may be interested in the social dynamics captured by $G$.

## 1.3.5 Multistage

*Multistage* sampling from finite populations is the case when the ultimate sample of elements is selected in several steps, and the sampling units are different at every step. For example, sampling of some (but not all the) hospitalised patients from an initial sample of hospitals corresponds to two-stage sampling.

A relational database is organised as a number of tables of rows and columns. Each row (or record) in a table has a unique *key* for identification. Each table represents a distinct type of entity, whose columns contain the associated values (or attributes). The keys of one table can appear as attributes in other tables.

Let $U$ be the union of all the collections of keys across the tables in a relational database. Let $(ij) \in A$ if the key $j$ is an attribute in the table where $i$ is an identification key, and $i$ and $j$ refer to different types of entities. The graph $G = (U, A)$ represents then the structure of the relational database, and selecting records from the database by keys can be represented as multistage OP in $G$.

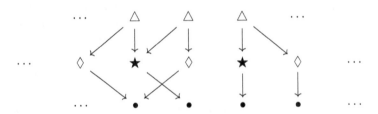

**Figure 1.5**  Relational database of entities △, ★, ◊ and •

Figure 1.5 illustrates a graph $G$ representing the structure of a payment-account-holder relational database containing 4 types of entity (the node set $U$): bank transaction (△), payer bank account (★), receiver bank account (◊), and bank account holder (•). Let payer account and receiver account be two attributes of the table of transactions, such that edges exist from △ to ★ or ◊ in the graph. Let holder be an attribute of the table of payer accounts and the table of receiver accounts, such that edges exist from ★ or ◊ to •. The edge set $A$ can also be divided into four parts: from transactions to payer accounts (△★), from transactions to receiver accounts (△◊), from payer accounts to holders (★•), and from receiver accounts to holders (◊•).

Given a sample of transactions $s_0$, apply first the incident OP to the edges from transactions to receiver accounts, then from receiver accounts to holders. This can be viewed as a 3-stage OP in $G$, where holders are the ultimate sampling units via the primary sampling units transactions and secondary sampling units receiver accounts. Subsampling of edges is possible at the 2nd and 3rd stages.

### 1.3.6 Multilayer

In a *multipartite* graph, the node set $U$ is partitioned into subsets, where edges only exist between nodes in different parts but not any nodes in the same part. In a *multilayer* graph, each edge in $A$ is of a specific type (or dimension), where edges can exist between any pair of nodes in $U$. Multipartite or multilayer graphs can be given as valued graphs, where the partition of nodes can be defined by a value associated with each node and that of edges by a value associated with each edge.

The graph of the relational database of payment-account-holder in Figure 1.5 can either be characterised as four-partite or four-dimensional. However, suppose additional edges are introduced, where $(ij), (ji) \in A$ if the two accounts $i$ and $j$ are affiliated with the same bank, and $(ij), (ji) \in A$ if the two holders $i$ and $j$ are mutual acquaintances. The graph is then multilayer and no longer simply multipartite.

Given a sample of transactions $s_0$, let the 1st-wave incident OP be applied to the edges from transactions to receiver accounts, and let the 2nd-wave incident OP be applied to the edges from receiver accounts to holders. In addition, let the incident OP be applied to the receiver accounts based on edges between them, and let the incident OP be applied to the holders based on edges between them. One may refer to the OP as *multistage-multilayer*, since it is also applied to edges between the same type of entities.

The OP is simply *multilayer* if it does not involve different types of nodes at all. For example, let $U$ consist of individuals. Let $A$ be two-dimensional, with edges between kins or colleagues. Separate incident observation applied to either type of edges is a multilayer OP. While multilayer OP resembles stratified sampling, in separate layers of edges rather than separate layers of units, multistage-multilayer OP in graphs is quite different to stratified multistage sampling from finite populations.

# 1.4 SAMPLE GRAPH, SAMPLING METHOD AND SAMPLING STRATEGY

## 1.4.1 Definition of sample graph

Let $G = (U, A)$ be a *population graph*. A method of sampling from $G$ has two parts:

- select an initial sample of nodes $s_0 \subset U$;

- given $s_0$, apply a specified OP making use of the edges in $A$.

The subgraph selected (or observed) accordingly is a *sample graph* from $G$, consisting of a node sample and an edge sample, which is given by

$$G_s = (U_s, A_s) \tag{1.3}$$

$$A_s = A \cap s_{\text{ref}} \qquad \text{and} \qquad U_s = s \cup \text{Inc}(A_s)$$

where $s$ is the *seed sample*, to which the OP has been applied under the given method of graph sampling.

The definition (1.3) includes the situation, where $s_0$ is given as the nodes that are incident to a sample of edges directly selected from $A$. It is then possible for the OP to specify that no additional edges need to be sampled, in which case $A_s$ contains the directly selected edges and $s = \emptyset$.

We do not consider separately sampling from graphs or valued graphs. Generally, a graph sampling method may depend on the values associated with $G$, and the values associated with the sample graph $G_s$ are observed together with $G_s$.

## 1.4.2 Basic graph sampling methods

*Depth-first* or *breadth-first* are two basic options for graph traversal or search, which are based on the incident OP. For instance, let ♠ mark the starting point and ★ the point of interest for search in Figure 1.6, where the progress of a depth-first or breath-first search algorithm is illustrated on the left or right side, respectively.

Adopting strictly probabilistic rules can turn these algorithms into basic graph sampling methods.

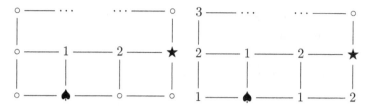

Depth-first (left): if possible go up, right, down *or* left
Breadth-first (right): if possible, go up, right, down *and* left

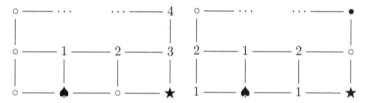

**Figure 1.6** Illustration of basic search options in graphs

For instance, from the current node $i$, one can select *randomly* one out-edge from $A_{i+}$, which would turn depth-first search into random walk sampling from $G$. The starting point ♠ may or may not be selected with a known probability. Or, one can choose the starting point ♠ *randomly* from all the nodes, which would turn breadth-first search into snowball sampling from $G$.

One can use an initial sample $s_0$ as multiple starting points. Simple random sampling (SRS) with or without replacement, Bernoulli sampling, and Poisson sampling are some basic methods for initial sampling of $s_0$.

A sampling design of $s_0$ is *symmetric* if $\pi_M = \Pr(M \subseteq s_0)$ for $M \subseteq U$ only depends on $|M|$ but is a constant of $M$ otherwise, for all $1 \leq |M| \leq N$. SRS with or without replacement and Bernoulli sampling are all symmetric designs. SRS without replacement is the only symmetric design with fixed sample size of distinct elements.

### 1.4.3 Graph sampling strategy

Given a population graph, $G = (U, A)$, representing a population of units, $U$, and the connections or links between them, $A$, one may be interested in the structure of the connections, or the links may provide effectively access to the part of population that is the primary target. Either way, graph sampling provides a valid statistical approach to study real graphs. Analogously to sampling from finite populations, it is based on exploring the variation over all possible sample graphs, or subgraphs, which can be taken from the given population graph by a specified method of sampling.

Let the target of interest be a graph total $\theta$ (or parameter $\mu$). A *graph sampling strategy* consists of a graph sampling method and an accompanying estimator. The properties of a strategy are evaluated over hypothetically repeated sampling from the *same* population graph. Inference with respect to the associated *known* sampling probabilities is said to be *design-based*, irrespectively of the unknown properties of the population graph.

## BIBLIOGRAPHIC NOTES

Neyman (1934) lays down the theoretical foundation of probability sampling from finite populations and outlines the concept of sampling strategy. Horvitz and Thompson (1952) developed the basic theory of HT-estimation. Cochran (1977) gives an account of the most common finite-population sampling techniques. A number of unconventional sampling methods are described in Thompson (2012), from either design or model-based perspective.

The adjacency matrix $\mathcal{A}$ (of $a_{ij}$) is defined to be symmetric for undirected graphs. Denote by $\mathcal{D}$ the diagonal matrix of degrees. The Laplacian matrix $\mathcal{D} - \mathcal{A}$ is of central interest in Spectral Graph Theory (e.g. Chung, 1997).

Newman (2010) gives a comprehensive introduction to network analysis and many graph characteristics of interest. For a statistical approach to graph problems, one may choose to model the valued graph as a random realisation; see e.g. Goldenberg et al. (2010) for a survey of statistical network models. Or, one may choose to exploit the variation over the possible sample graphs taken from a given real graph. Graph sampling theory deals with valued graphs under the latter perspective.

In the tradition of computer science, one may be concerned with situations where the graph is in principle known but is too large or dynamic to be fully processed or stored practically (e.g. Leskovec and Faloutsos, 2006; Hu and Lau, 2013). Graph sampling provides then a possible approach.

For an early reference of graph sampling theory, Goodman (1961) studies the motif of mutual best friendships in a digraph where $a_{i+} = 1$ for all $i \in U$.

Ove Frank has many contributions to graph sampling. See e.g. Frank (1977c, 1979, 1980b, 1981, 2011) for his own summary. For instance, Frank studies samples of nodes (Frank, 1971; 1977c; 1994), dyads (Frank, 1971; 1977a; 1977b; 1979) or triads (Frank, 1971; 1979), where a sample of motifs from the population of motifs is conceived in analogy to a sample $s$ from the population $U$. Or, Frank discusses various graph characteristics, such as order (Frank, 1971; 1977c; 1994), size (Frank, 1971; 1977a; 1977b; 1979), degree (Frank, 1971; 1980a), connectedness (Frank, 1971; 1978).

Zhang and Patone (2017) propose a general definition of sample graph, where sampling is completely driven by the incident edges, whether the OP is multiwave, adaptive or random. This provides a complete analogy between sample graph as a sub-graph and sample as a sub-population, where different kinds of motifs can be observed in a sample graph. The definition of sample graph (1.3) slightly generalises the definition in Zhang and Patone (2017), allowing the OP to include random jumps to and from isolated nodes.

Zhang (2021) provides a short introduction to graph sampling, mainly by examples. Several figures and tables in Zhang (2021) are reproduced or rearranged in this book.

# CHAPTER 2

# Bipartite incidence graph sampling and weighting

For distinction, denote by $\mathcal{B} = (F, \Omega; H)$ a simple digraph, where $(F, \Omega)$ form a bipartition of the node set $U = F \cup \Omega$, and each edge in $H$ points from one node in $F$ to another in $\Omega$. We consider sampling from a given population graph $\mathcal{B}$ in this chapter. In the later chapters, $\mathcal{B}$ will be used for constructing strategies of graph sampling from arbitrary population graph $G$.

## 2.1 BIPARTITE INCIDENCE GRAPH SAMPLING

*Bipartite incidence graph sampling (BIGS)* from $\mathcal{B} = (F, \Omega; H)$ is given by applying the incident OP to an initial (or seed) sample $s_0 \subset F$. By (1.3), this yields the sample graph

$$\mathcal{B}_s = (s_0, \Omega_s; H_s) \qquad (2.1)$$

with bipartite $U_s = s_0 \cup \Omega_s$, where $\Omega_s = \alpha(s_0)$ is a sample of nodes from $\Omega$, and $s_{\text{ref}} = s_0 \times \Omega$ such that $H_s = H \cap s_{\text{ref}}$.

We use $i, j, h$ to enumerate $F$, which are said to be the *sampling units* and have known initial sample inclusion probabilities, such as $\pi_i$ or $\pi_{ij}$. We use $\kappa, \ell$ to enumerate $\Omega$, which are referred to as the *study units*, such that the graph total of interest is defined over $\Omega$ by (1.2). Each edge $(i\kappa)$ in $H$ is given the meaning that $\kappa$ is observed whenever $i$ is sampled. Generally, $F$ is assumed to be known, while both $H$ and $\Omega$ may be unknown.

Inference under BIGS is based on the sample inclusion probabilities. Despite the known initial-sample inclusion probabilities, the study-sample inclusion probabilities $\pi_{(\kappa)} = \Pr(\kappa \in \Omega_s)$ and $\pi_{(\kappa\ell)} = \Pr(\kappa \in \Omega_s, \ell \in \Omega_s)$ can only be calculated provided the knowledge of their *ancestors* (or multiplicity) in $F$, which are

$$\{\beta_\kappa : \kappa \in \Omega_s\} \quad \text{and} \quad \beta(\Omega_s) = \bigcup_{\kappa \in \Omega_s} \beta_\kappa$$

and are referred to as the *ancestry knowledge*. In this chapter, we assume the ancestry knowledge is given in addition to the sample graph. Notice that we only need the ancestry knowledge of the actual sample $\Omega_s$. Moreover, as necessary conditions for design-based inference, we assume

$$\Omega = \alpha(F) = \bigcup_{i \in F} \alpha_i \quad \text{and} \quad \Pr(s_0 \cap \beta_\kappa \neq \emptyset) > 0, \ \forall \kappa \in \Omega$$

Consider e.g. BIGS from Figure 2.1, where $F = \{i_1, i_2, i_3, i_4\}$, $\Omega = \{\kappa_1, \kappa_2, \kappa_3\}$ and $H = \{(i_1\kappa_1), (i_2\kappa_1), (i_2\kappa_2), (i_3\kappa_3)\}$. Suppose $s_0 = \{i_1, i_3\}$, then $\Omega_s = \{\kappa_1, \kappa_3\}$, $H_s = \{(i_1\kappa_1), (i_3\kappa_3)\}$ and $\mathcal{B}_s = (s_0, \Omega_s; H_s)$ by (2.1). Moreover, the ancestry knowledge of $\Omega_s$ consists of $\beta_{\kappa_1} = \{i_1, i_2\}$, $\beta_{\kappa_3} = \{i_3\}$ and $\beta(\Omega_s) \setminus s_0 = \{i_2\}$.

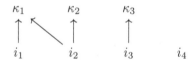

**Figure 2.1**  An illustration of BIGS

Figure 2.2 illustrates *finite-population sampling* as special cases of BIGS. For element sampling, let $F = \Omega = U$, where $(i\kappa) \in H$ iff $i$ and $\kappa$ refer to the same element. For cluster sampling, let $F$ consist of the $m_F$ clusters in the population and $\Omega = U = \bigcup_{i \in F} U_i$ the elements, where $(i\kappa) \in H$ iff element $\kappa \in U_i$. The mapping from $F$ to $\Omega$ is one-one for element sampling or one-many for cluster sampling, so that the ancestry knowledge is guaranteed in either case. The mapping can be many-one or many-many under BIGS, such that the ancestry knowledge must be required explicitly.

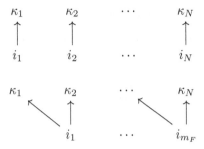

**Figure 2.2** Finite population element sampling (top) or cluster sampling (bottom) as BIGS

## 2.2 INCIDENCE WEIGHTING ESTIMATOR

Given $\mathcal{B}_s$, assign the *incidence weights* $\{W_{i\kappa} : (i\kappa) \in H_s\}$ to the sample edges. The *incidence weighting estimator (IWE)* of the graph total (1.2) is given by

$$\hat{\theta} = \sum_{(i\kappa) \in H_s} W_{i\kappa} \frac{y_\kappa}{\pi_i} \qquad (2.2)$$

Notice that, under BIGS, we have

$$\pi_{(i\kappa)} := \Pr\big((i\kappa) \in H_s\big) = \Pr(i \in s_0) = E(\delta_i) = \pi_i$$

where $\delta_i = 1$ or $0$ indicates if $i \in s_0$ or not. The definition (2.2) allows $W_{i\kappa}$ to vary with $\mathcal{B}_s$. Replacing $\pi_i$ by $\pi_{(i\kappa)}$ in (2.2) would allow the OP to involve subsampling among $A_{i+}$ for each $i \in s_0$, in which case the results below for unbiased IWE and its associated variance can easily be rephrased accordingly.

**Theorem 2.1.** *The IWE by (2.2) is unbiased for $\theta$ by (1.2) if, for each $\kappa \in \Omega$, we have*

$$\sum_{i \in \beta_\kappa} E(W_{i\kappa} | \delta_i = 1) = 1 \qquad (2.3)$$

*Proof.* The expectation of $\hat{\theta}$ over repeated sampling is

$$E(\hat{\theta}) = \sum_{i \in F} \frac{E(\delta_i)}{\pi_i} E\big(\sum_{\kappa \in \alpha_i} W_{i\kappa} y_\kappa | \delta_i = 1\big)$$

$$= \sum_{\kappa \in \Omega} y_\kappa \sum_{i \in \beta_\kappa} E(W_{i\kappa} | \delta_i = 1) = \theta$$

given $\pi_{(i\kappa)} = \pi_i$ under BIGS and $\sum_{i\in\beta_\kappa} E(W_{i\kappa}|\delta_i = 1) = 1$. □

**Corollary 2.1.** *If the incidence weights are constant of sampling, denoted by $\omega_{i\kappa}$, then the IWE is unbiased for $\theta$ by (1.2) if, for each $\kappa \in \Omega$, we have*

$$\sum_{i\in\beta_\kappa} \omega_{i\kappa} = 1 \tag{2.4}$$

**Proposition 2.1.** *The BIGS variance of an unbiased IWE is*

$$V(\hat{\theta}) = \sum_{\kappa\in\Omega}\sum_{\ell\in\Omega}(\Delta_{\kappa\ell} - 1)y_\kappa y_\ell \tag{2.5}$$

*where*

$$\Delta_{\kappa\ell} = \sum_{i\in\beta_\kappa}\sum_{j\in\beta_\ell}\frac{\pi_{ij}}{\pi_i\pi_j}E(W_{i\kappa}W_{j\ell}|\delta_i\delta_j = 1)$$

*Proof.* Given unbiased $\hat{\theta}$, we have $V(\hat{\theta}) = E(\hat{\theta}^2) - \theta^2$, where

$$E(\hat{\theta}^2) = \sum_{\kappa\in\Omega}\sum_{\ell\in\Omega}y_\kappa y_\ell \sum_{i\in\beta_\kappa}\sum_{j\in\beta_\ell}E\left(\frac{\delta_i\delta_j}{\pi_i\pi_j}W_{i\kappa}W_{j\ell}\right)$$

$$= \sum_{\kappa\in\Omega}\sum_{\ell\in\Omega}y_\kappa y_\ell \sum_{i\in\beta_\kappa}\sum_{j\in\beta_\ell}\frac{\pi_{ij}}{\pi_i\pi_j}E(W_{i\kappa}W_{j\ell}|\delta_i\delta_j = 1)$$

The result follows on substituting $\theta^2 = (\sum_{\kappa\in\Omega} y_\kappa)^2$. □

## 2.2.1  HT-estimator (HTE)

Under BIGS, we have

$$\pi_{(\kappa)} = \Pr(\kappa \in \Omega_s) = 1 - \bar{\pi}_{\beta_\kappa} = 1 - \Pr(\beta_\kappa \cap s_0 = \emptyset)$$

$$\pi_{(\kappa\ell)} = \Pr(\kappa \in \Omega_s, \ell \in \Omega_s) = 1 - \bar{\pi}_{\beta_\kappa} - \bar{\pi}_{\beta_\ell} + \bar{\pi}_{\beta_\kappa\cup\beta_\ell}$$

where $\bar{\pi}_{\beta_\kappa}$, $\bar{\pi}_{\beta_\ell}$ and $\bar{\pi}_{\beta_\kappa\cup\beta_\ell}$ are the initial sample exclusion probabilities. The HT-estimator (HTE) is given by

$$\hat{\theta}_y = \sum_{\kappa\in\Omega_s}\frac{y_\kappa}{\pi_{(\kappa)}} \tag{2.6}$$

and its sampling variance is given by

$$V(\hat{\theta}_y) = \sum_{\kappa\in\Omega}\sum_{\ell\in\Omega}\left(\frac{\pi_{(\kappa\ell)}}{\pi_{(\kappa)}\pi_{(\ell)}} - 1\right)y_\kappa y_\ell$$

Provided $\pi_{(\kappa\ell)} > 0$ for all $(\kappa\ell)$, an unbiased variance estimator is

$$\hat{V}(\hat{\theta}_y) = \sum_{\kappa \in \Omega_s} \sum_{\ell \in \Omega_s} \left( \frac{1}{\pi_{(\kappa)}\pi_{(\ell)}} - \frac{1}{\pi_{(\kappa\ell)}} \right) y_\kappa y_\ell$$

The HTE is a special case of the IWE, where the incidence weights $W_{i\kappa}$ satisfy

$$\sum_{i \in s_0 \cap \beta_\kappa} \frac{W_{i\kappa}}{\pi_i} = \frac{1}{\pi_{(\kappa)}} \qquad (2.7)$$

which are not constant of sampling if $|\beta_\kappa| > 1$, depending on how $s_0$ intersects $\beta_\kappa$. For BIGS from Figure 2.1, we have

$$W_{i_2\kappa_2} = \frac{\pi_{i_2}}{\pi_{(\kappa_2)}} \qquad \text{and} \qquad W_{i_3\kappa_3} = \frac{\pi_{i_3}}{\pi_{(\kappa_3)}}$$

by (2.7), since $|\beta_{\kappa_2}| = |\beta_{\kappa_3}| = 1$ in $\mathcal{B}$. Moreover,

$$\begin{cases} W_{i_1\kappa_1} = \frac{\pi_{i_1}}{\pi_{(\kappa_1)}} & \text{if } i_1 \in s_0, \ i_2 \notin s_0 \\ W_{i_2\kappa_1} = \frac{\pi_{i_2}}{\pi_{(\kappa_1)}} & \text{if } i_1 \notin s_0, \ i_2 \in s_0 \\ (W_{i_1\kappa_1}, W_{i_2\kappa_1}) = \left(b\frac{\pi_{i_1}}{\pi_{(\kappa_1)}}, \ (1-b)\frac{\pi_{i_2}}{\pi_{(\kappa_1)}}\right) & \text{if } i_1, i_2 \in s_0 \end{cases}$$

for any value $b$, since the contribution of any $y_\kappa$ to (2.2) is in terms of the coefficient $\sum_{i \in s_0 \cap \beta_\kappa} W_{i\kappa}/\pi_i$. A numerical illustration of $W_{i\kappa_1}$ is given in Table 2.1 in Section 2.3.

To see that the weights given by (2.7) satisfy the condition (2.3) generally, let $\phi_{s_\kappa}$ be the probability that the *initial sample intersection* is

$$s_\kappa = s_0 \cap \beta_\kappa$$

for $\kappa \in \Omega$, where

$$\pi_{(\kappa)} = \sum_{s_\kappa} \phi_{s_\kappa}$$

over all the distinct $s_\kappa$. Given (2.7) for any $\kappa \in \Omega$, we have

$$\sum_{i \in \beta_\kappa} E(W_{i\kappa}|\delta_i = 1) = \sum_{i \in \beta_\kappa} \sum_{s_\kappa \ni i} \frac{\phi_{s_\kappa}}{\pi_i} W_{i\kappa}$$

$$= \sum_{s_\kappa} \phi_{s_\kappa} \sum_{i \in s_\kappa} \frac{W_{i\kappa}}{\pi_i} = \sum_{s_\kappa} \frac{\phi_{s_\kappa}}{\pi_{(\kappa)}} = 1$$

Similarly, it can be shown that the term $\Delta_{\kappa\ell}$ in (2.5) reduces to $\pi_{(\kappa\ell)}/\pi_{(\kappa)}\pi_{(\ell)}$ given (2.7) and (2.3).

## 2.2.2 Hansen-Hurwitz (HH) type estimator

The Hansen-Hurwitz (HH) type estimator uses weights $\omega_{i\kappa}$ that are constant of sampling, such that it can be given by

$$\hat{\theta}_z = \sum_{i \in s_0} \frac{z_i}{\pi_i} \qquad \text{and} \qquad z_i = \sum_{\kappa \in \alpha_i} \omega_{i\kappa} y_\kappa \qquad (2.8)$$

where $z_i$ is a constant, $\forall i \in F$. It is unbiased given (2.4), since

$$\sum_{i \in F} \sum_{\kappa \in \alpha_i} \omega_{i\kappa} y_\kappa = \sum_{\kappa \in \Omega} y_\kappa \sum_{i \in \beta_\kappa} \omega_{i\kappa} = \sum_{\kappa \in \Omega} y_\kappa$$

in which case its sampling variance following (2.5) is given by

$$V(\hat{\theta}_z) = \sum_{i \in F} \sum_{j \in F} \left( \frac{\pi_{ij}}{\pi_i \pi_j} - 1 \right) z_i z_j$$

For BIGS from Figure 2.1, we have $\omega_{i_2 \kappa_2} = \omega_{i_3 \kappa_3} = 1$ and $\omega_{i_1 \kappa_1} + \omega_{i_2 \kappa_1} = 1$. Thus, the expression (2.8) defines actually a family of estimators, depending on the choice of $\omega_{i\kappa}$. In particular, the equal weights

$$\omega_{i\kappa} = |\beta_\kappa|^{-1}$$

are referred to as the *multiplicity weights*, and the corresponding IWE is the *multiplicity estimator*, denoted by $\hat{\theta}_{z\beta}$.

## 2.2.3 Priority-rule estimator

For each $\kappa \in \Omega_s$, let $I_{i\kappa} = 1$ if $i = \min\left(s_0 \cap \beta_\kappa\right)$ and 0 otherwise, depending on if $i$ happens to be enumerated first in $F$ among the sampled ancestors of $\kappa$. The *priority-rule estimator* is based on the subset $\{(i\kappa) : I_{i\kappa} = 1, (i\kappa) \in H_s\}$ and is given by

$$\hat{\theta}_p = \sum_{(i\kappa) \in H_s} \frac{I_{i\kappa} \omega_{i\kappa} y_\kappa}{p_{i\kappa} \pi_i} \qquad (2.9)$$

(Birnbaum and Sirken, 1965), where

$$p_{i\kappa} := \Pr\left(I_{i\kappa} = 1 | (i\kappa) \in H_s\right) = \Pr\left(I_{i\kappa} = 1 | \delta_i = 1\right)$$

is the conditional probability that $(i\kappa)$ is prioritised given it is included in $H_s$, and $\omega_{i\kappa}$ is the multiplicity weight. Clearly, other priority rules or $\omega_{i\kappa}$ are possible.

One can easily recognise $\hat{\theta}_p$ as a special case of (2.2) with

$$W_{i\kappa} = I_{i\kappa}\omega_{i\kappa}/p_{i\kappa}$$

It satisfies the unbiasedness condition (2.3), provided $p_{i\kappa} > 0$ for all $(i\kappa) \in H$, in which case $E(W_{i\kappa}|\delta_i = 1) = \omega_{i\kappa}$. Its variance follows from (2.5), where

$$\Delta_{\kappa\ell} = \sum_{i\in\beta_\kappa} \sum_{j\in\beta_\ell} \frac{\pi_{ij}p_{i\kappa,j\ell}}{\pi_i\pi_j p_{i\kappa}p_{j\ell}} \omega_{i\kappa}\omega_{j\ell}$$

and $p_{i\kappa,j\ell} := \Pr(I_{i\kappa}I_{j\ell} = 1|\delta_i\delta_j = 1)$, such that

$$V(\hat{\theta}_p) = \sum_{(i\kappa)\in H} \sum_{(j\ell)\in H} \left(\frac{\pi_{ij}p_{i\kappa,j\ell}}{\pi_i\pi_j p_{i\kappa}p_{j\ell}} - 1\right)\omega_{i\kappa}\omega_{j\ell}y_\kappa y_\ell$$

because $\sum_{i\in\beta_\kappa}\omega_{i\kappa} = 1$ for any $\kappa \in \Omega$. Provided $\pi_{ij} > 0$ for all $i, j \in F$, an unbiased variance estimator is

$$\hat{V}(\hat{\theta}_p) = \sum_{(i\kappa)\in H_s} \sum_{(j\ell)\in H_s} \left(\frac{\pi_{ij}p_{i\kappa,j\ell}}{\pi_i\pi_j p_{i\kappa}p_{j\ell}} - 1\right)\frac{\omega_{i\kappa}\omega_{j\ell}}{\pi_{ij}}y_\kappa y_\ell$$

The priority probabilities $p_{i\kappa}$ and $p_{i\kappa,j\ell}$ depend on both the priority rule and the initial sampling design of $s_0$. For the priority rule $\min(s_0 \cap \beta_\kappa)$, let $d_{i(\kappa)} = \sum_{j\in F:j<i}\mathbb{I}((j\kappa) \in H)$ be the number of nodes with higher priority than $i$ for each $\kappa \in \Omega$ and $i \in \beta_\kappa$. Suppose initial SRS with $m = |s_0|$. Let $m_F = |F|$. We have

$$p_{i\kappa} = \binom{m_F - 1 - d_{i(\kappa)}}{m - 1} / \binom{m_F - 1}{m - 1}$$

and

$$p_{i\kappa,j\ell} = \begin{cases} p_{i\kappa} & \text{if } \kappa = \ell, i = j \\ 0 & \text{if } \kappa = \ell, i \neq j \\ \binom{m_F-1-d_{i(\kappa,\ell)}}{m-1}/\binom{m_F-1}{m-1} & \text{if } \kappa \neq \ell, i = j \end{cases}$$

whereas for $\kappa \neq \ell$ and $i \neq j$, $p_{i\kappa,j\ell}$ is given by

$$\begin{cases} \binom{m_F-2-d_{i(\kappa),j(\ell)}}{m-2}/\binom{m_F-2}{m-2} & \text{if } |\beta_\kappa^i \cap \{j\}| + |\beta_\ell^j \cap \{i\}| = 0 \\ 0 & \text{if } |\beta_\kappa^i \cap \{j\}| + |\beta_\ell^j \cap \{i\}| > 0 \end{cases}$$

where $d_{i(\kappa,\ell)} = |\beta_\kappa^i \cup \beta_\ell^i|$ and $d_{i(\kappa),j(\ell)} = |\beta_\kappa^i \cup \beta_\ell^j|$ and $\beta_\kappa^i$ is the subset ancestors of $\kappa$ with higher priority than $i$.

The priority rule selects only one sample edge for each study unit in $\Omega_s$ for the purpose of estimation. The sample graph $\mathcal{B}_s$ includes all the edges incident to every node in $s_0$ by definition. There is a possibility that an edge in $H$ can be sampled but never prioritised, in which case $\hat{\theta}_p$ would be biased. For an extreme example, suppose a study unit $\kappa$ is adjacent to all the nodes in $F$, then $(i\kappa)$ can never be prioritised, where $i$ is the last node in $F$, according to the priority rule $\min(s_0 \cap \beta_\kappa)$ as long as $|s_0| > 1$. Generally, $\hat{\theta}_p$ is biased under this priority rule, provided there exists at least one $\kappa$ in $\Omega$ with $|\beta_\kappa| > 1$, where

$$\Pr(|s_0 \cap \beta_\kappa| > 1 \mid \kappa \in \Omega_s) = 1$$

such that $(i\kappa)$ with $i = \max(\beta_\kappa)$ has no chance of being prioritised when it is in $s_0$. The probability above depends on the ordering of nodes in $F$, as well as the initial sample size. Given any ordering of $F$, as the initial sample size increases, it is possible for $\hat{\theta}_p$ to become unstable or biased.

## 2.3   RAO-BLACKWELLISATION

Given an unbiased estimator, its conditional expectation given the minimal sufficient statistic is also an unbiased estimator, known as the Rao-Blackwell (RB) method. If the minimal sufficient statistic is also complete, then the RB-estimator is the unique minimum-variance unbiased estimator (UMVUE).

The minimal sufficient statistic for BIGS is $\{(\kappa, y_\kappa) : \kappa \in \Omega_s\}$ or simply $\Omega_s$ as long as one keeps in mind that the $y$-values are associated constants. However, it is not complete, because $y_\kappa$ can be an arbitrary constant. It follows that the RB method generally does not lead to UMVUE for BIGS.

Applying the RB method to an unbiased IWE $\hat{\theta}$ yields

$$\hat{\theta}_{RB} = E(\hat{\theta}|\Omega_s)$$

which has a smaller variance than $\hat{\theta}$ if $V(\hat{\theta}|\Omega_s)$ is positive. Since the HTE $\hat{\theta}_y$ is fixed conditional on $\Omega_s$, it will remain unchanged, and it is only possible to use the RB method to improve the efficiency of a non-HT estimator.

For an illustration using Figure 2.1, consider first BIGS given $|s_0| = 1$. There are 4 distinct initial samples leading to 4 distinct $\Omega_s$, such that $V(\hat{\theta}|\Omega_s) = 0$ and $\hat{\theta}_{RB} = \hat{\theta}$ for any unbiased IWE. Next, given $|s_0| = 2$, there are 6 different initial samples leading to 5 distinct $\Omega_s$, where $s_0 = \{i_1, i_2\}$ and $s'_0 = \{i_2, i_4\}$ yield the same $\Omega_s = \{\kappa_1, \kappa_2\}$, so that $\hat{\theta}_{RB} \neq \hat{\theta}$ if $\hat{\theta}(s_0) \neq \hat{\theta}(s'_0)$. For the HH-type estimator $\hat{\theta}_z$ by (2.8), we have

$$\hat{\theta}_z(s_0) = \frac{\omega_{i_1 \kappa_1}}{\pi_{i_1}} y_{\kappa_1} + \frac{\omega_{i_2 \kappa_1}}{\pi_{i_2}} y_{\kappa_1} + \frac{\omega_{i_2 \kappa_2}}{\pi_{i_2}} y_{\kappa_2}$$

$$\hat{\theta}_z(s'_0) = \frac{\omega_{i_2 \kappa_1}}{\pi_{i_2}} y_{\kappa_1} + \frac{\omega_{i_2 \kappa_2}}{\pi_{i_2}} y_{\kappa_2}$$

$$\hat{\theta}_{zRB} = \frac{p(s_0)}{p(s_0) + p(s'_0)} \cdot \frac{\omega_{i_1 \kappa_1}}{\pi_{i_1}} y_{\kappa_1} + \frac{\omega_{i_2 \kappa_1}}{\pi_{i_2}} y_{\kappa_1} + \frac{\omega_{i_2 \kappa_2}}{\pi_{i_2}} y_{\kappa_2}$$

The RB method may be infeasible if the conditional sample space of $s_0$ given $\Omega_s$ is too large or if the distribution $p(s_0)$ is not fully specified. Nevertheless, the reasoning of Rao-Blackwellisation can offer some insights, as discussed below.

### 2.3.1 HT-type estimator

The HTE is based on sample-dependent incidence weights $W_{i\kappa}$ that satisfy the constraint (2.7). More generally, let

$$\eta_{s_\kappa} = \pi_{(\kappa)} \sum_{i \in s_\kappa} \frac{W_{i\kappa}}{\pi_i}$$

To satisfy the condition (2.3) for any $\kappa \in \Omega$, the weights must be such that

$$\sum_{s_\kappa} \phi_{s_\kappa} \eta_{s_\kappa} = \pi_{(\kappa)}$$

The HTE is the special case of $\eta_{s_\kappa} \equiv 1$. It is possible to assign $\eta_{s_\kappa}$ that differs from 1 for different initial sample intersections subject to this restriction. Any such estimator may be referred to as an *HT-type estimator*. However, applying the RB method to an HT-type estimator would recover the HTE, because

$$E\left( \sum_{\kappa \in \Omega_s} \sum_{i \in s_\kappa} \frac{W_{i\kappa}}{\pi_i} y_\kappa | \Omega_s \right) = \sum_{\kappa \in \Omega_s} y_\kappa E\left( \frac{\eta_{s_\kappa}}{\pi_{(\kappa)}} | \kappa \in \Omega_s \right)$$

$$= \sum_{\kappa \in \Omega_s} \frac{y_\kappa}{\pi_{(\kappa)}} \sum_{s_\kappa} \frac{\phi_{s_\kappa}}{\pi_{(\kappa)}} \eta_{s_\kappa} = \sum_{\kappa \in \Omega_s} \frac{y_\kappa}{\pi_{(\kappa)}}$$

**Table 2.1**   Weight $W_{i\kappa_1}$ by $s_{\kappa_1}$ under BIGS from Figure 2.1 given SRS with $|s_0| = 2$

| $s_{\kappa_1}$ | HTE | | | HT-type, an example | | |
|---|---|---|---|---|---|---|
| | $\{i_1, i_2\}$ | $\{i_1\}$ | $\{i_2\}$ | $\{i_1, i_2\}$ | $\{i_1\}$ | $\{i_2\}$ |
| $W_{i_1\kappa_1}$ | $b(3/5)$ | $3/5$ | $-$ | $b$ | $2/5$ | $-$ |
| $W_{i_2\kappa_1}$ | $(1-b)(3/5)$ | $-$ | $3/5$ | $1-b$ | $-$ | $3/5$ |

Table 2.1 illustrates numerically the HTE-weights $W_{i\kappa_1}$ under BIGS from Figure 2.1 given SRS of $s_0$ with $|s_0| = 2$, where $\pi_i \equiv 1/2$ and $\pi_{(\kappa_1)} = 5/6$ such that $\pi_i/\pi_{(\kappa_1)} \equiv 3/5$. Given $\eta_{s_\kappa} \equiv 1$, we have $W_{i_1\kappa_1} = 3/5$ if $s_{\kappa_1} = \{i_1\}$, $W_{i_2\kappa_1} = 3/5$ if $s_{\kappa_1} = \{i_2\}$, and $W_{i_1\kappa_1} + W_{i_2\kappa_1} = 3/5$ if $s_{\kappa_1} = \{i_1, i_2\}$.

Next, as an example of HT-type estimator, let $\eta_{\{i_1\}} = 2/3$ and $W_{i_1\kappa_1} = 2/5$ if $s_{\kappa_1} = \{i_1\}$, and $\eta_{\{i_2\}} = 1$ and $W_{i_2\kappa_1} = 3/5$ if $s_{\kappa_1} = \{i_2\}$. Since $\phi_{s_{\kappa_1}} = 1/3$ if $s_{\kappa_1} = \{i_1\}$ or $\{i_2\}$ and $\phi_{s_{\kappa_1}} = 1/6$ if $s_{\kappa_1} = \{i_1, i_2\}$, we must have

$$\frac{1}{6}\,\eta_{\{i_1, i_2\}} = \frac{5}{6} - \left(\frac{1}{3}\right)\left(\frac{2}{3}\right) - \frac{1}{3} = \frac{5}{18}$$

$$W_{i_1\kappa_1} + W_{i_2\kappa_1} = \frac{3}{5}\,\eta_{\{i_1, i_2\}} = 1$$

These weights are given in the right half of Table 2.1.

## 2.3.2   HH-type estimator

Consider the special case of many-one mapping from $F$ to $\Omega$, where $|\alpha_i| = a_{i+} \equiv 1$. Compare $\hat{\theta}_y$ and $\hat{\theta}_z$ given $|s_0| = 1$. Let $p_i$ and $p_{(\kappa)} = \sum_{i \in \beta_\kappa} p_i$ be the respective *selection* probabilities of $i \in F$ and $\kappa \in \Omega$. We have $p_{ij} = p_i$ if $i = j$ and $0$ if $i \neq j$, and $p_{(\kappa\ell)} = p_{(\kappa)}$ if $\kappa = \ell$ and $0$ if otherwise, so that

$$V\Big(\sum_{i \in s_0} \frac{z_i}{p_i}\Big) - V\Big(\sum_{\kappa \in \Omega_s} \frac{y_\kappa}{p_{(\kappa)}}\Big) = \sum_{\kappa \in \Omega}\Big(\sum_{i \in \beta_\kappa} \frac{\omega_{i\kappa}^2}{p_i} - \frac{1}{p_{(\kappa)}}\Big)y_\kappa^2 = 0$$

only if $\omega_{i\kappa} \equiv p_i/p_{(\kappa)}$ for $i \in \beta_\kappa$ and $\hat{\theta}_z = \hat{\theta}_{zRB}$. The variance of any other $\hat{\theta}_z$ would be larger, as long as $\omega_{i\kappa}/p_i$ is not a constant over $\beta_\kappa$, because

$$E\left(\frac{z_i}{p_i}\Big|\kappa \in \Omega_s\right) = \sum_{i \in \beta_\kappa} \frac{p_i w_{i\kappa} y_\kappa}{p_{(\kappa)} p_i} = \frac{y_\kappa}{p_{(\kappa)}}$$

$$V\left(\frac{z_i}{p_i}\Big|\kappa \in \Omega_s\right) = y_\kappa^2 V\left(\frac{\omega_{i\kappa}}{p_i}\Big|\kappa \in \Omega_s\right) > 0$$

Since the same holds for with-replacement sampling of $s_0$ from $F$, where $|s_0| > 1$ and $|\alpha_i| \equiv 1$, it suggests the choice $\omega_{i\kappa} \propto \pi_i$ under sampling of $s_0$ without replacement, provided $\pi_{ij} \approx \pi_i \pi_j$ and $\pi_{(\kappa\ell)} \approx \pi_{(\kappa)} \pi_{(\ell)}$. This can make $z_i/\pi_i$ more similar to each other over $F$, which is advantageous in terms of the mean squared error of $\hat{\theta}_z$, evaluated under the sampling design *and* a population model of $z_i$ (Godambe and Joshi, 1965, Theorem 6.2).

To make $z_i/\pi_i$ more similar to each other without the restriction $|\alpha_i| \equiv 1$, one may consider setting $\omega_{i\kappa} < \omega_{j\kappa}$ if $a_{i+} > a_{j+}$, despite $\pi_i = \pi_j$, because there are more study units contributing to $z_i$ than $z_j$. Thus, it may be reasonable to consider the *probability and inverse-degree adjusted (PIDA)* weights, which are given by

$$\omega_{i\kappa} \propto \pi_i a_{i+}^{-\gamma} \tag{2.10}$$

subjected to (2.4), where $\gamma > 0$ is a chosen tuning constant.

Denote by $\hat{\theta}_{z\alpha\gamma}$ the corresponding PIDA-IWE. The multiplicity estimator $\hat{\theta}_{z\beta}$ becomes a special case of $\hat{\theta}_{z\alpha\gamma}$ if $\gamma = 0$ and $\pi_i$ is constant over $F$.

To apply the weights (2.10) with $\gamma \neq 0$, one needs to know $a_{i+}$ for all $i \in \beta(\Omega_s) \setminus s_0$. For instance, let $F$ be the parents, $\Omega$ the children and $H$ the connections between parents and children. One would need to obtain the number of children for the out-of-sample parents in $\beta(\Omega_s) \setminus s_0$ as well.

## 2.4 ILLUSTRATIONS

Let $y_\kappa \equiv 1$ for $\kappa \in \Omega$, such that the graph total by (1.2) is $\theta = |\Omega|$. BIGS can be used to estimate $\theta$ in many situations.

- Let $F$ be the population and $\Omega$ the cases in an epidemiological study, where $(i\kappa) \in H$ if person $i$ is in-contact with case $\kappa$, including when $i$ and $\kappa$ refer to the same person.

- Let $F$ be the Twitter accounts and $\Omega$ the followers. Let $\alpha_i$ be the followers of $i \in F$.

- Let $F$ be the products available in an online market place and $\Omega$ the buying customers. Let $\alpha_i$ be those who have purchased the product $i \in F$.

Consider the following IWEs given SRS of $s_0$ from $F$:

- the HTE $\hat{\theta}_y$ by (2.6);

- the HH-type estimator $\hat{\theta}_{za\gamma}$ by (2.10), where $\hat{\theta}_{za0} = \hat{\theta}_{z\beta}$;

- the priority-rule estimator $\hat{\theta}_p$ by (2.9) with multiplicity weights $\omega_{i\kappa}$ and priority rule $\min(s_0 \cap \beta_\kappa)$, where the sampling units in $F$ are arranged in random, ascending or descending order of $a_{i+}$, yielding $\hat{\theta}_{pR}$, $\hat{\theta}_{pA}$ and $\hat{\theta}_{pD}$, respectively.

### 2.4.1 A numerical example

Consider BIGS from Figure 2.3 with SRS of $s_0$ and $m = |s_0| = 2$.

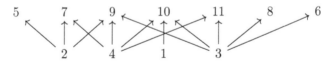

**Figure 2.3** An example of BIGS from Patone (2020)

The PIDA weights and the corresponding $z$-values are given in Table 2.2. Given $\pi_i \equiv |s_0|/m_F = 0.5$ with $m_F = |F|$, we have

$$V(\hat{\theta}_z) = m_F^2 \left( \frac{1}{m} - \frac{1}{m_F} \right) S_z^2$$

$$S_z^2 = \sum_{i \in F} \frac{(z_i - \bar{z})^2}{m_F - 1} \quad \text{and} \quad \bar{z} = \sum_{i \in F} \frac{z_i}{m_F}$$

The inverse-degree adjustment $a_{i+}^{-\gamma}$ can heavily reduce the variance compared to the multiplicity weights.

Table 2.2 PIDA weights and $z$-value given $\gamma$ under BIGS from Figure 2.3

| $\gamma$ | $\omega_{1,10}$ | $\omega_{2,5}$ | $\omega_{2,7}$ | $\omega_{2,9}$ | $\omega_{3,10}$ | $\omega_{3,8}$ | $\omega_{3,11}$ | $\omega_{3,9}$ | $\omega_{3,6}$ | $\omega_{4,7}$ | $\omega_{4,10}$ | $\omega_{4,11}$ | $\omega_{4,9}$ |
|---|---|---|---|---|---|---|---|---|---|---|---|---|---|
| 0 | 0.33 | 1 | 0.50 | 0.33 | 0.33 | 1 | 0.5 | 0.33 | 1 | 0.5 | 0.33 | 0.5 | 0.33 |
| 1 | 0.69 | 1 | 0.57 | 0.43 | 0.14 | 1 | 0.44 | 0.26 | 1 | 0.43 | 0.17 | 0.56 | 0.32 |
| 2 | 0.90 | 1 | 0.64 | 0.52 | 0.04 | 1 | 0.39 | 0.19 | 1 | 0.36 | 0.06 | 0.61 | 0.29 |
| 3 | 0.98 | 1 | 0.70 | 0.61 | 0.007 | 1 | 0.34 | 0.14 | 1 | 0.30 | 0.013 | 0.66 | 0.25 |

| $\gamma$ | $z_1$ | $z_2$ | $z_3$ | $z_4$ | $S_z^2$ |
|---|---|---|---|---|---|
| 0 | 0.33 | 1.83 | 3.17 | 1.67 | 1.34 |
| 1 | 0.69 | 2.00 | 2.83 | 1.48 | 0.81 |
| 2 | 0.91 | 2.16 | 2.61 | 1.32 | 0.60 |
| 3 | 0.98 | 2.31 | 2.48 | 1.23 | 0.57 |

**Table 2.3** Incidence weights $W_{i\kappa}$ by priority rule given $\tilde{F} = \{2, 4, 1, 3\}$ in Figure 2.3

| $s_0$ | $W_{1,10}$ | $W_{2,5}$ | $W_{2,7}$ | $W_{2,9}$ | $W_{3,10}$ | $W_{3,8}$ | $W_{3,11}$ | $W_{3,9}$ | $W_{3,6}$ | $W_{4,7}$ | $W_{4,10}$ | $W_{4,11}$ | $W_{4,9}$ |
|---|---|---|---|---|---|---|---|---|---|---|---|---|---|
| {1,2} | 0.33 | 1 | 0.5 | 0.33 | - | - | - | - | - | - | - | - | - |
| {1,3} | 0.33 | - | - | - | 0 | 1 | 0.5 | 0.5 | 1 | - | - | - | - |
| {1,4} | 0.33 | - | - | - | - | - | - | - | - | 0.75 | 0 | 0.75 | 1 |
| {2,3} | - | 1 | 0.5 | 0.33 | 0.5 | 1 | 0.5 | 0 | 1 | - | - | - | - |
| {2,4} | - | 1 | 0.5 | 0.33 | - | - | - | - | - | 0 | 1 | 0.75 | 0 |
| {3,4} | - | - | - | - | 0.5 | 1 | 0.5 | 0.5 | 1 | 0.75 | 0 | 0 | 0 |

The incidence weights $W_{i\kappa}$ for the priority-rule estimator (2.9) vary with the ordered frame, denoted by $\tilde{F}$, where $\tilde{F} = \{2, 4, 1, 3\}$ in Figure 2.3. Whereas we would have $\tilde{F} = \{3, 4, 2, 1\}$ if the nodes are arranged in the descending order of $a_{i+}$, or $\tilde{F} = \{1, 2, 4, 3\}$ in the ascending order of $a_{i+}$.

The incidence weights given $\tilde{F} = \{2, 4, 1, 3\}$ are shown in Table 2.3. For instance, given $s_0 = \{1, 2\}$, we have $\Omega_s = \{10, 5, 7, 9\}$, and all the sample edges in $H_s$ are prioritised, since $|s_0 \cap \beta_\kappa| = 1$ for each $\kappa \in \Omega_s$. Whereas given $s_0 = \{1, 3\}$, we have $\Omega_s = \{10, 9, 11, 8, 6\}$, and all the sample edges are prioritised except that from 3 to 10, since $s_0 \cap \beta_{10} = \{1, 3\}$ and the edge from 1 to 10 is prioritised.

**Table 2.4** Variance of IWEs under BIGS from Figure 2.3 given SRS with $|s_0| = 2$

| $\hat{\theta}_{z\alpha 0}$ | $\hat{\theta}_{z\alpha 1}$ | $\hat{\theta}_{z\alpha 2}$ | $\hat{\theta}_{z\alpha 3}$ | $\hat{\theta}_{pR}$ | $\hat{\theta}_{pD}$ | $\hat{\theta}_{pA}$ | $\hat{\theta}_y$ |
|---|---|---|---|---|---|---|---|
| 5.37 | 3.25 | 2.41 | 2.28 | 3.06 | 2.55 | 6.32 | 3.98 |

Table 2.4 gives the sampling variances, which vary from 2.28 at the lowest to 6.32 at the highest. The best HH-type estimator here has variance 2.28, and the best priority-rule estimator has variance 2.55, both of which are more efficient than the HTE.

## 2.4.2  A simulation study

Two graphs $\mathcal{B} = (F, \Omega; H)$ and $\mathcal{B}' = (F, \Omega; H')$ are constructed for the same $F$ and $\Omega$, where $|F| = 54$ and $|\Omega| = 310$. The two edge sets have the same number of edges, $|H| = |H'| = 1200$, but different distributions of $a_{i+}$ over $F$, as shown in Figure 2.4. The distribution is relatively uniform over a small range of values in $\mathcal{B}$, but much more skewed and asymmetric in $\mathcal{B}'$.

Consider SRS of $s_0$ with varying $m = |s_0| = 2, ..., 53$. Table 2.5 gives the relative efficiency (RE) of six other estimators against the HTE, for a selected set of initial sample sizes, each based on 10000 simulations of BIGS from either $\mathcal{B}$ or $\mathcal{B}'$. All the results are significant with respect to the simulation error.

It can be seen in Table 2.5 that the priority-rule estimator is always dominated by some HH-type estimators in this simulation study. Moreover, from details omitted here, it is observed that all the three estimators $\hat{\theta}_{pR}$, $\hat{\theta}_{pA}$ and $\hat{\theta}_{pD}$ become biased given large

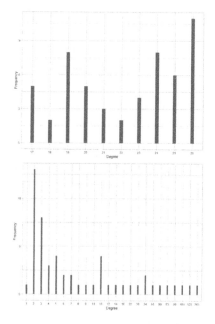

**Figure 2.4**   Distribution of $a_{i+}$ in $\mathcal{B}$ (top) and $\mathcal{B}'$ (bottom)

enough initial sample size $m$, which happens at $m = 45$ for $\mathcal{B}$ where the maximum in-degree $a_{+\kappa}$ is 10 over $\Omega$, and $m = 46$ for $\mathcal{B}'$ where the maximum in-degree $a_{+\kappa}$ is 9. Moreover, although the variance of any $\hat{\theta}_p$ initially decreases as $m$ increases, the variance starts to increase with $m$ once the latter is larger than a threshold value, somewhere between 10 and 30 in these simulations, so that the performance of $\hat{\theta}_p$ can deteriorate as the initial sample size increases long before it becomes biased.

The sampling variance of the priority-rule estimator $\hat{\theta}_p$ is also affected by the ordering of the nodes in $F$. The variance tends to be lowest when $F$ is arranged in the descending order of $a_{i+}$, as long as the variance is decreasing with $m$, whereas the ascending ordering tends to yield the largest variance. Without prioritisation, the value $z_i$ is a constant of sampling given $\omega_{i\kappa}$. Due to the randomness induced by the priority rule, $z_i$ varies over different samples. A node with large $a_{i+}$ has a large range of possible $z_i$ values. Placing such a node towards the end of the ordering tends to increase the sample variance of $\{z_i : i \in s_0\}$ due to prioritisation, compared to when the same node is placed towards the beginning of the

Table 2.5  RE of IWE against HTE from 10000 simulations

| m | $\hat{\theta}_{z\alpha0}$ | $\hat{\theta}_{z\alpha1}$ | $\hat{\theta}_{z\alpha2}$ | $\hat{\theta}_{pR}$ | $\hat{\theta}_{pA}$ | $\hat{\theta}_{pD}$ |
|---|---|---|---|---|---|---|
| | | | | $\mathcal{B}$ | | |
| 5 | 0.96 | 0.55 | 0.49 | 0.80 | 1.43 | 0.68 |
| 11 | 0.95 | 0.55 | 0.48 | 0.97 | 2.57 | 0.84 |
| 17 | 0.99 | 0.57 | 0.51 | 2.34 | 4.98 | 2.57 |
| 29 | 1.31 | 0.75 | 0.67 | 26.7 | 30.1 | 33.2 |
| m | $\hat{\theta}_{z\alpha0}$ | $\hat{\theta}_{z\alpha1}$ | $\hat{\theta}_{z\alpha2}$ | $\hat{\theta}_{pR}$ | $\hat{\theta}_{pA}$ | $\hat{\theta}_{pD}$ |
| | | | | $\mathcal{B}'$ | | |
| 5 | 1.22 | 0.23 | 0.18 | 0.97 | 1.16 | 0.83 |
| 11 | 1.74 | 0.33 | 0.25 | 0.89 | 1.54 | 0.45 |
| 17 | 2.67 | 0.51 | 0.39 | 0.82 | 2.30 | 0.24 |
| 29 | 7.96 | 1.54 | 1.17 | 12.0 | 12.1 | 29.3 |

ordering, because the rule $\min(s_0 \cap \beta_\kappa)$ favours the node in front of the other ancestors. This is a reason why the descending ordering of $a_{i+}$ may work better than the ascending ordering. However, one may not know $\{a_{i+} : i \in F\}$ in advance, in which case applying $\hat{\theta}_p$ given *ad hoc* ordering of $F$ can be a haphazard business.

Given initial SRS, the different HH-type estimators based on the PIDA weights (2.10) differ only with respect to the use of $a_{i+}$ via the choice of $\gamma$. The multiplicity estimator $\hat{\theta}_{z\alpha0}$ is the least efficient here, especially for $\mathcal{B}'$ where the distribution of $a_{i+}$ is more skewed. The differences between the other two estimators $\hat{\theta}_{z\alpha1}$ and $\hat{\theta}_{z\alpha2}$ are relatively small, compared to their differences to $\hat{\theta}_{z\alpha0}$, so that a non-optimal choice of $\gamma \neq 0$ is less critical than simply setting $\gamma = 0$. These results suggest that the extra effort that may be required to obtain $a_{i+}$ is worth considering in applications.

Finally, both $\hat{\theta}_{z\alpha1}$ and $\hat{\theta}_{z\alpha2}$ are more efficient than the HT-estimator $\hat{\theta}_y$ when $m$ is small, whereas $\hat{\theta}_y$ improves more quickly as $m$ becomes larger, especially for $\mathcal{B}'$. Generally, the difference between the two sampling variances can be given as

$$V(\hat{\theta}_z) - V(\hat{\theta}_y) = \sum_{\kappa \in \Omega} \sum_{\ell \in \Omega} \left( \sum_{i \in \beta_\kappa} \sum_{j \in \beta_\ell} \frac{\pi_{ij}}{\pi_i \pi_j} \omega_{i\kappa} \omega_{j\ell} - \frac{\pi_{(\kappa\ell)}}{\pi_{(\kappa)} \pi_{(\ell)}} \right) y_\kappa y_\ell$$

The RE between the two depends on the interplay of the sample inclusion probabilities of the nodes in $F$ and $\Omega$, which is complex as it depends on the population edge set $H$. The matter is not simplified given initial SRS, where we simply have $\pi_i = m/m_F$ and $\pi_{ij} = m(m-1)/m_F(m_F - 1)$ for any $i \neq j \in F$.

## BIBLIOGRAPHIC NOTES

Birnbaum and Sirken (1965) study the situation where patients are sampled indirectly via the "medical sources" from which they receive treatment, which may be physicians in individual practice, clinics or hospitals. Insofar as a patient may be treated at more than one source, the patients are not nested in the sources like elements in clusters. The knowledge of multiplicity (of the sampled patients) is needed in order to calculate the HTE. Indirect sampling as such is naturally represented as BIGS, where $F$ consists of the sources and $\Omega$ the patients, with an edge between a source and each of its patients. The ancestry knowledge amounts to $\beta_\kappa$ for each sampled patient $\kappa \in \Omega_s$.

Birnbaum and Sirken (1965) do not cast the problem in terms of graph sampling. They identify the condition (2.4) for an unbiased HH-type estimator (2.8), although they only use the multiplicity weights. Adaptions of the multiplicity estimator to other settings of indirect, network sampling are e.g. considered by Sirken (1970), Sirken and Levy (1974), Sirken (2004, 2005) and Lavallee (2007). A modified multiplicity estimator is considered for adaptive cluster sampling (Thompson, 1990; 1991).

The priority-rule estimator is the third estimator of Birnbaum and Sirken (1965), who do not provide an expression of its sampling variance but indicate that it is unwieldy. The estimator seems to have vanished from the literature since then. Patone and Zhang (2020) provide the variance of $\hat{\theta}_p$ under SRS of $s_0$ and the priority rule $\min(s_0 \cap \beta_\kappa)$. Moreover, they show that the application of the priority-rule estimator may be a haphazard business, unless one is able to control the interplay between the ordering of $F$ and the priority-rule, and the estimator may become biased as the initial sample size increases and behave erratically long before that.

Patone and Zhang (2020) formulate the unifying class of IWE for BIGS and identify the general condition (2.3) for unbiased estimation, which allows for sample-dependent weights $W_{i\kappa}$ beyond the constant weights $\omega_{i\kappa}$ and the associated condition (2.4). In particular, the HTE is shown to be a special case of IWE.

The RB method (Rao, 1945; Blackwell, 1947) does not figure prominently in the theory of finite population sampling, where it does not lead to the UMVUE, and the classic HTE is unchanged. A notable exception is the works on adaptive sampling, for which modifications of the HTE and the multiplicity estimator are used;

see e.g. Thompson (1990; 1991; 2006b), Dryver and Thompson (2005), Vincent and Thompson (2017).

The numerical example in Section 2.4.1 is given in Patone (2020), and the simulation study in Section 2.4.2 is taken from Patone and Zhang (2020).

# CHAPTER 3

# Strategy BIGS-IWE

Suitably constructed, the strategy BIGS-IWE is applicable to many sampling situations, whether or not these are originally cast as graph problems.

## 3.1 APPLICABILITY

For any given method of graph sampling from $G = (U, A)$, which starts from an initial sample $s_0$ and yields the sample graph $G_s$ by (1.3), let $F \subseteq U$ be the *initial sampling frame*, consisting of the nodes in $U$, which have positive inclusion probabilities in $s_0$. Let $\theta$ be the graph total given by (1.2), where each motif $\kappa$ in $\Omega$ involves the nodes $M(\kappa)$ or simply $M$ as long as the context is clear that $M$ and $\kappa$ correspond to each other. Let $\Omega_s$, $\Omega_s \subseteq \Omega$, be the motifs that are observed in the sample graph $G_s$. As a condition for applying the strategy BIGS-IWE, we assume the given graph sampling method is such that, for any $\kappa \in \Omega$, we have

$$F_\kappa = \{i \in F : \Pr(\kappa \in \Omega_s | i \in s_0) = 1\} \neq \emptyset \qquad (3.1)$$

The nodes $F_\kappa$ are said to be the *ancestors* of $\kappa$ with respect to the given graph sampling method. In other words, for any $\kappa \in \Omega$, there exists a nonempty ancestor set $F_\kappa \subseteq F$, such that we have $\kappa \in \Omega_s$ whenever $F_\kappa \cap s_0 \neq \emptyset$.

Note that graph sampling based on the induced OP cannot satisfy (3.1) generally. For instance, let $\Omega = A$ contain the edges of $G$, where a motif $\kappa = (hj)$ is observed in $\Omega_s = A_s$ iff $(h, j) \in s_0$, because of the induced OP, such that generally we would have $\Pr(\kappa \in \Omega_s | i \in s_0) < 1$ for any $i \in F$, including when $i = h$ or $j$.

Given (3.1), define an *associated BIGS* from $\mathcal{B} = (F, \Omega; H)$ with

$$H = \bigcup_{\kappa \in \Omega} \beta_\kappa \times \kappa \quad \text{and} \quad \emptyset \neq \beta_\kappa \subseteq F_\kappa \tag{3.2}$$

where $\beta_\kappa$ is a *fixed nonempty subset* of $F_\kappa$ for $\kappa \in \Omega$, such that the sample graph $\mathcal{B}_s$ from $\mathcal{B}$ is given by (2.1), with the motif sample

$$\Omega_s(\mathcal{B}) = \{\kappa \in \Omega : s_0 \cap \beta_\kappa \neq \emptyset\}$$

Notice that a motif $\kappa$ may be in $\Omega_s$ but not $\Omega_s(\mathcal{B})$ if $F_\kappa \cap s_0 \neq \emptyset$ but $\beta_\kappa \cap s_0 = \emptyset$. For an illustration, let $F = U$ and let $\Omega$ contain all the 2-stars in an undirected graph $G$, such as the motif $\kappa$ of $M = \{i_1, i_2, i_3\}$ in Figure 3.1. Consider 2-wave incident OP given $s_0 \subset F$. We have $F_\kappa = \{i_1, i_2, i_3, i_4\}$ by (3.1). Suppose we set $\beta_\kappa = M$ in (3.2). Then we would have $\kappa \in \Omega_s$ and $\kappa \notin \Omega_s(\mathcal{B})$, if $i_4 \in s_0$ while $M \cap s_0 = \emptyset$.

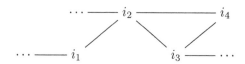

**Figure 3.1**   An illustration for associated BIGS by (3.2)

Whether the choice $\beta_\kappa$ can be made *a priopri* for any $\kappa \in \Omega$ or based on the actual sample graph $G_s$ for any $\kappa \in \Omega_s$, one can apply the IWE (2.2) for BIGS by (3.2) to obtain unbiased estimation of $\theta$ given by (1.2), where the IWE can be written as

$$\hat{\theta} = \sum_{\kappa \in \Omega} \mathbb{I}\big(\kappa \in \Omega_s(\mathcal{B})\big) \sum_{i \in \beta_\kappa \cap s_0} W_{i\kappa} \frac{y_\kappa}{\pi_i}$$

Note that $\beta_\kappa$ and $\beta_{\kappa'}$ can be fixed separately for any two different motifs $\kappa \neq \kappa' \in \Omega$.

**Theorem 3.1.** *For any graph sampling method from $G = (U, A)$ defined by (1.3) and subjected to (3.1), an associated BIGS-IWE strategy defined for $\mathcal{B}$ given by (3.2) and IWE subjected to (2.3) is unbiased for $\theta$ defined by (1.2).*

*Proof.* Given (3.1) for sampling from $G$ and nonempty $\beta_\kappa$ in $\mathcal{B}$ given by (3.2), every motif in $\Omega$ has a positive probability of being included in $\Omega_s(\mathcal{B})$ with respect to the associated BIGS. The IWE (2.2) can be specified whether $\beta_\kappa$ is fixed before sampling or based on the actual sample graph $G_s$ for any $\kappa \in \Omega_s(\mathcal{B})$. Subjected to (2.3), the IWE is unbiased for $\theta$ by Theorem 2.1 over repeated sampling from $G$. □

## 3.2 NETWORK SAMPLING

*Network sampling* is the case if each network is sampled iff any of the nodes in the network is sampled. Take sampling of siblings as an example here, where the siblings report each other such that a network of siblings are all sampled iff any one of them is sampled. Consider two settings of initial sampling: of individuals, or of households that leads to the household members indirectly.

Let $F = U$ consist of a population of relevant individuals from which the initial sample $s_0$ can be selected. Let $(ij),(ji) \in A$ iff $i$ and $j$ are siblings of each other. Each sibling network is a clique in $G = (U, A)$. Examples of sibling networks of order 1, 2 and 3 are given in Figure 3.2. Let $\Omega$ consist of all the sibling networks in $G$. For each $\kappa \in \Omega$ with nodes $M$, we have simply $F_\kappa = M$ by (3.1) under network sampling.

**Figure 3.2**   Network sampling of siblings directly

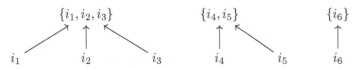

**Figure 3.3**   Network sampling from Figure 3.2 as BIGS

For an associated BIGS, we can simply let $\beta_\kappa = F_\kappa$ in (3.2), which is unknown in advance but observed in every $G_s$ where $\kappa$

is observed. Figure 3.3 provides an illustration for the motifs in Figure 3.2, where each motif is shown as its node set. Regarding the IWE, this is the situation with $|\alpha_i| \equiv 1$ in Section 2.3.

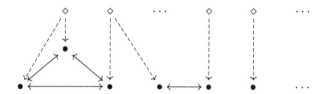

**Figure 3.4**   Network sampling of siblings (•) via households (◇)

**Figure 3.5**   Network sampling from Figure 3.4 as BIGS

Consider initial sampling of households, where the siblings may belong to different households, some of which are outside of the initial sample. Figure 3.4 illustrates the situation as graph sampling from $G = (U, A)$, where the node set $U$ contains all the households (◇) and siblings (•), with examples of sibling networks of order 1, 2 and 3. The edges from ◇ to • determine the multiplicity of access from households to siblings, while the edges between two •-nodes stand for the sibling relationship as in Figure 3.2.

Let $F$ be the initial sampling frame consisting of the ◇-nodes. Let $\Omega$ consist of all the sibling networks as before. For each $\kappa \in \Omega$, $F_\kappa$ now consists of the ◇-nodes that point to any node in $M$. Figure 3.5 gives the associated BIGS for the motifs in Figure 3.4, where each motif is shown as its node set and $\beta_\kappa = F_\kappa$ in (3.2). This associated BIGS is possible if the network sampling OP in $G$ is formally 3-wave incident *reciprocal*, so that $F_\kappa$ is fully observed in every $G_s$ where $\kappa$ is observed.

Network sampling in $G$ is also possible if the OP is formally 3-wave incident, where one observes the ancestors that are selected in $s_0$ for any $\kappa$ in $G_s$, denoted by $F_{\kappa|s} = s_0 \cap F_\kappa$, but not any ancestor outside of $s_0$. One can let $\beta_\kappa = F_{\kappa|s}$ in (3.2) for an associated BIGS. For any sibling network $\kappa$ with $|F_\kappa| > 1$, the choice $\beta_\kappa$ would thus depend on how the first time $\kappa$ is observed over hypothetically repeated sampling from $G$. Whereas we have $\beta_\kappa = F_{\kappa|s} \equiv F_\kappa$ for any $\kappa$ with $|F_\kappa| = 1$. Fixing $\beta_\kappa$ in this way for every $\kappa \in \Omega$, the strategy BIGS-IWE is unbiased for $\theta$ given by (1.2).

## 3.3 LINE-INTERCEPT SAMPLING

Line-intercept sampling (LIS) is a method of sampling habitats in a given area, where habitat is sampled if a chosen line segment transects it. The habitats are of irregular shapes, such as animal tracks or forests. In a simple setting, each transect line is given by selecting randomly a position along a fixed *baseline* that traverses the whole study area in the direction perpendicular to the baseline. In a more general setting, a point is randomly selected on the map, and an angle is randomly chosen, yielding a line segment of fixed length or transecting the whole area in the chosen direction. Repetition of either procedure generates an IID sample of lines.

### 3.3.1 Associated BIGS

Becker (1991) gives an example of baseline-LIS, where the aim is to estimate the total number of wolverines, denoted by $\theta$, in the boxed area sketched in Figure 3.6.

Four systematic samples A, B, C and D are drawn on the baseline that is equally divided into 3 sections of length 12 miles each. Following the 12 selected lines *and* any wolverine track that intercepts them yields the four observed tracks, denoted by $\kappa = 1, ..., 4$ and heuristically indicated by the dashed lines in Figure 3.6. Let $y_\kappa$ be the associated number of wolverines, and $L_\kappa$ the length of the projection of $\kappa$ on the baseline. From top to bottom and left to right, we observe $(y_1, L_1) = (1, 5.25)$, $(y_2, L_2) = (2, 7.5)$, $(y_3, L_3) = (2, 2.4)$ and $(y_4, L_4) = (1, 7.05)$.

Given the observed tracks, partition the baseline into *projection segments*, each with length $x_i$, for $i = 1, ..., m_F^*$ and $m_F^* = 7$. The probability that the $i$-th projection segment is selected under systematic sampling here is $p_i = x_i/12$ where, from left to right, $x_1$ refers to the overlapping projection of $\kappa = 1$ and 2, $x_2$ the projection of $\kappa = 2$ that does not overlap with $\kappa = 1$, $x_3$ the distance between projections of $\kappa = 2$ and 3, $x_4$ the projection of $\kappa = 3$, $x_5$ the distance between projections of $\kappa = 3$ and 4, $x_6$ the projection of $\kappa = 4$, and $x_7$ the distance between $\kappa = 4$ and right-hand border.

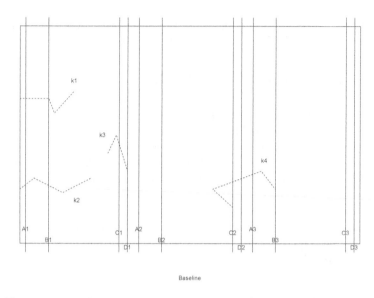

Baseline

**Figure 3.6** LIS with 4 systematic samples (A, B, C, D) of 3 positions each, heuristic sketches of wolverine tracks

Let $s_r$ contain the selected projection segments on the $r$-th draw, and $\Omega_r$ the tracks observed from $s_r$. In this case, we have $s_1 = s_2 = \{1, 5, 6\}$, yielding $\Omega_1 = \Omega_2 = \{1, 2, 4\}$ on the first two draws A and B, and $s_3 = s_4 = \{4, 6, 7\}$, yielding $\Omega_3 = \Omega_4 = \{3, 4\}$ on the last two draws C and D. The distinct projection segments selected over all the draws are $s = \bigcup_{r=1}^{4} s_r = \{1, 4, 5, 6, 7\}$, and the distinct tracks are $\Omega_s = \bigcup_{r=1}^{4} \Omega_r = \{1, 2, 3, 4\}$.

$$\begin{array}{cccc} \kappa = 1 & \kappa = 2 & \kappa = 3 & \kappa = 4 \\ \uparrow & \nearrow \uparrow & \uparrow & \uparrow \\ i = 1 & i = 2 \quad i = 3 & i = 4 \quad i = 5 & i = 6 \quad i = 7 \end{array}$$

**Figure 3.7** $\mathcal{B}^*$ constructed from realised line-intercept samples

Let $F^* = \{i : i = 1, ..., m_F^*\}$ be the projection segments that are constructed from the actual samples $s$ and $\Omega_s$. Let $\mathcal{B}^* = (F^*, \Omega_s; H^*)$ be given in Figure 3.7, where $(i\kappa) \in H^*$ if $\kappa$ is observed if the $i$-th projection segment is selected. Denote by $\beta_\kappa^*$ the ancestors of $\kappa$ in $\mathcal{B}^*$, where $\beta_1^* = \{1\}$, $\beta_2^* = \{1, 2\}$, $\beta_3^* = \{4\}$ and $\beta_4^* = \{6\}$.

Let $\Omega = \{1, ..., \kappa, ..., |\Omega|\}$ be all the wolverine tracks in the area, where $|\Omega| \geq 4$. Let $F = \{1, ..., i, ..., m_F\}$ be the projection segments constructed from $\Omega$. For any $i \in F$ and $\kappa \in \Omega$, let $(i\kappa) \in H$ if $\kappa$ intercepts any line that originates from the $i$-th projection segment, yielding $\mathcal{B} = (F, \Omega; H)$.

Only $\mathcal{B}^*$ can be constructed but not $\mathcal{B}$ in applications. The two are not the same generally, in that one needs to further partition the projection segments of $F^*$ in $F$ based on $\Omega$ if there exist unobserved tracks in $\Omega \setminus \Omega_s$. For instance, suppose there is a track that can only be intercepted from the 7th projection segment in $F^*$ and the track does not reach the right-hand border, then this projection segment would be partitioned into 3 segments in $F$ and $(F, H)$ would differ from $(F^*, H^*)$.

Under LIS, field observation along a line has an actual width of detectability. Dividing the baseline accordingly would yield a known sampling frame $F'$ of the detectability partitions. Define $\mathcal{B}' = (F', \Omega; H')$, where $(i\kappa) \in H'$ provided $\kappa$ intercepts any line that originates from the $i$-th detectability partition. Clearly, applying the strategy BIGS-IWE with respect to $\mathcal{B}'$ would yield unbiased estimation of $\theta$.

Now, as along as the unit of detectability is negligible in scale compared to the baseline, one can assume the elements of $F'$ to be nested in those of $F^*$ (or $F$), such that the selection probability of each $\kappa$ under BIGS from $\mathcal{B}'$ can be correctly calculated using $\mathcal{B}^*$ (or $\mathcal{B}$). Thus, the strategy BIGS-IWE defined for $\mathcal{B}'$ can be applied using the observed $\mathcal{B}^*$, just as when $\mathcal{B}$ were known.

## 3.3.2 HTE

When the position of the transect line is selected randomly over the entire baseline, the selection probability of track $\kappa$ on each draw is $p_{(\kappa)} = \sum_{i \in \beta_\kappa^*} p_i$, calculated with respect to $\mathcal{B}^*$, where $p_i \propto x_i$ and $\sum_{i \in F^*} p_i = 1$. The inclusion probability of $\kappa \in \Omega_s$ is given by $\pi_{(\kappa)} = 1 - (1 - p_{(\kappa)})^4$, one minus the probability that $\kappa$ is not selected on any of the 4 draws. Denote by $p_{(\kappa \cup \ell)} = \sum_{i \in \beta_\kappa^* \cup \beta_\ell^*} p_i$ the probability of selecting either $\kappa$ or $\ell$ on a given draw. The second-order inclusion probability of $\kappa \neq \ell \in \Omega_s$ is given by

$$\pi_{(\kappa\ell)} = 1 - \Big( \Pr(\kappa \notin \Omega_s) + \Pr(\ell \notin \Omega_s) - \Pr(\kappa \notin \Omega_s, \ell \notin \Omega_s) \Big)$$
$$= \pi_{(\kappa)} + \pi_{(\ell)} - 1 + \big(1 - p_{(\kappa\cup\ell)}\big)^4$$

Notice that the computation of $p_{(\kappa\cup\ell)}$ contains some additional details when a systematic sample of 3 positions are drawn each time (Thompson, 2012). We have

$$p_{(1)} = 0.4375 \qquad p_{(2)} = 0.625 \qquad p_{(3)} = 0.2 \qquad p_{(4)} = 0.5875$$
$$\pi_{(1)} = 0.90 \qquad \pi_{(2)} = 0.98 \qquad \pi_{(3)} = 0.59 \qquad \pi_{(4)} = 0.97$$
$$\pi_{(12)} = 0.90 \qquad \pi_{(13)} = 0.51 \qquad \pi_{(14)} = 0.88$$
$$\pi_{(23)} = 0.57 \qquad \pi_{(24)} = 0.95 \qquad \pi_{(34)} = 0.59$$

Accordingly, the HTE and its estimated variance are

$$\hat{\theta}_y = \sum_{\kappa \in \Omega_s} \frac{y_\kappa}{\pi_{(\kappa)}} = 7.57 \qquad \text{and} \qquad \hat{V}(\hat{\theta}_y) = 5.27$$

### 3.3.3  HH-type estimators

An unbiased estimator of $\theta$ from the $r$-th draw is

$$\tau_r = \sum_{\kappa \in \Omega_r} \frac{y_\kappa}{p_{(\kappa)}}$$

where $\tau_1 = \tau_2 = 7.1878$ and $\tau_3 = \tau_4 = 11.7021$. The Hansen-Hurwitz (HH) estimator as the mean of $\tau_r$ over all the draws and its estimated variance are

$$\hat{\theta}_{HH} = \frac{1}{4} \sum_{r=1}^{4} \tau_r = 9.44 \qquad \text{and} \qquad \hat{V}(\hat{\theta}_{HH}) = 1.70$$

By the strategy BIGS-IWE using $\mathcal{B}^*$, the HH-type estimator (2.8) on the $r$-th draw is

$$\hat{\theta}_{z,r} = \sum_{i \in s_r} \frac{z_i}{p_i} \qquad \text{where} \qquad z_i = \sum_{\kappa \in \alpha_i^*} \omega_{i\kappa} y_\kappa$$

The multiplicity estimator $\hat{\theta}_{z\beta}$ uses $\omega_{11} = \omega_{43} = \omega_{64} = 1$ and $\omega_{12} = \omega_{22} = 0.5$. The resulting IWE and its estimated variance are

$$\hat{\theta}_{z\beta,1} = \hat{\theta}_{z\beta,2} = 6.2736 \qquad \hat{\theta}_{z\beta,3} = \hat{\theta}_{z\beta,4} = 11.7021$$
$$\hat{\theta}_{z\beta} = 8.99 \qquad \text{and} \qquad \hat{V}(\hat{\theta}_{z\beta}) = 2.46$$

The PIDA weights (2.10) with $\gamma = 0$ yields different $\omega_{i2} = p_i/p_{(2)}$ for $i \in \beta_2^*$ in particular, where $\omega_{12} = 0.7$ and $\omega_{22} = 0.3$. We obtain

$$\hat{\theta}_{z,r} = \sum_{i \in s_r} \frac{1}{p_i} \sum_{\kappa \in \alpha_i^*} \frac{p_i}{p_{(\kappa)}} y_\kappa = \sum_{\kappa \in \Omega_r} \frac{y_\kappa}{p_{(\kappa)}} = \tau_r$$

on the $r$-th draw, such that the HH-estimator is just the IWE $\hat{\theta}_{z\alpha 0}$. One can also use other values of $\gamma$ in (2.10). For instance, we have $\omega_{12} = 0.6226$ and $\omega_{22} = 0.3773$ given $\gamma = 0.5$. The resulting IWE $\hat{\theta}_{z\alpha.5}$ and its estimated variance are given by

$$\hat{\theta}_{z\alpha.5,1} = \hat{\theta}_{z\alpha.5,2} = 6.8341 \qquad \hat{\theta}_{z\alpha.5,3} = \hat{\theta}_{z\alpha.5,4} = 11.7021$$

$$\hat{\theta}_{z\alpha.5} = 9.27 \quad \text{and} \quad \hat{V}(\hat{\theta}_{z\alpha.5}) = 1.97$$

Setting $\gamma = 1.227$ would yield the multiplicity estimator $\hat{\theta}_{z\beta}$.

### 3.3.4 Remarks

All the results detailed above are summarised in Table 3.1. Neither the HTE $\hat{\theta}_y$ nor the multiplicity estimator $\hat{\theta}_{z\beta}$ is efficient here. Efficiency gains can be achieved using the PIDA weights (2.10). In this case, adjusting the equal weights by the selection probability while disregarding the degrees of the nodes in $F$ performs well, where $\hat{\theta}_{z\alpha 0}$ has the lowest estimated variance. Whether or not the true variance of $\hat{\theta}_{z\alpha 0}$ is smaller than some $\hat{\theta}_{z\alpha\gamma}$ with $\gamma \neq 0$, it seems that a reasonably efficient IWE can be obtained if one is able to choose the coefficient $\gamma$ in (2.10) appropriately.

Table 3.1   Illustrated results of BIGS-IWE using $\mathcal{B}^*$

| Estimate | $\hat{\theta}_y$ | $\hat{\theta}_{z\beta}$ | $\hat{\theta}_{z\alpha 0}$ | $\hat{\theta}_{z\alpha.5}$ |
|---|---|---|---|---|
| Total | 7.57 | 8.99 | 9.44 | 9.27 |
| Variance | 5.27 | 2.46 | 1.70 | 1.97 |

Given the systematic sampling design of the transect lines, the tracks $\{1, 2, 4\}$ can only be observed if a position is selected in the left part of 1st projection segment, which would result in $\{1, 5, 6\}$ as the sampled projection segments. The tracks $\{3, 4\}$ can only be observed if a position is selected in the 4th projection segment, which would result in $\{4, 6, 7\}$ as the sampled projection segments. Thus, applying the RB method would not change any IWE for BIGS from $\mathcal{B}^*$ in this case.

## 3.4 SAMPLING FROM RELATIONAL DATABASES

Let $G = (U, A)$ be the structure of a relational database, where $U$ contains the keys of all the tables. Suppose three types of entities: stock transaction $\{h, i, j, ...\}$, buyer $\{b, c, ...\}$, and company $\{\kappa, \ell, ...\}$. Let $(hb) \in A$ if transaction $h$ involves buyer $b$, where the mapping from transactions to buyers is many-to-one. Let $(b\kappa) \in A$ if buyer $b$ has ownership in company $\kappa$, where the mapping from buyers to companies is many-to-many.

### 3.4.1 Sampling

Given an initial sample $s_0$ from all the transactions $F$, suppose one traces first their buyers and then the companies in which these buyers have ownership. Let $\Omega$ consist of the companies that can be traced in this way. Let $\theta = \sum_{\kappa \in \Omega} y_\kappa$, where the $y$-values are unknown to start with, but can be collected for a given sample of companies, in order to estimate $\theta$. Suppose it is possible to identify the ancestors $F_\kappa$ for any $\kappa \in \Omega_s$ and the condition (3.1) is satisfied.

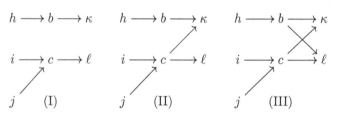

**Figure 3.8** Some configurations of relational database structure

Denote by $G$ the graph representing the database structure. Some configurations are given as (I) - (III) in Figure 3.8, where $\mathcal{B}_1 = (F, Q; H_1)$ and $\mathcal{B}_2 = (Q, \Omega; H_2)$ are fused together by $Q$ that contains all the buyers. Using configuration (II) to illustrate the notation, we have

$$\begin{cases} \alpha_h = \{b\}, \ \alpha_h^2 = \alpha(\alpha_h) = \{\kappa\} \\ \alpha_i = \{c\}, \ \alpha_i^2 = \alpha(\alpha_i) = \{\kappa, \ell\} \\ \alpha_j = \{c\}, \ \alpha_j^2 = \alpha(\alpha_j) = \{\kappa, \ell\} \end{cases}$$

and

$$\begin{cases} \beta_\kappa = \{b, c\}, \ \beta_\kappa^2 = \beta(\beta_\kappa) = \{h, i, j\} \\ \beta_\ell = \{c\}, \ \beta_\ell^2 = \beta(\beta_\ell) = \{i, j\} \end{cases}$$

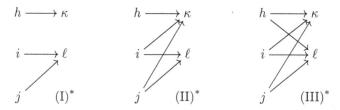

**Figure 3.9**  Associated BIGS corresponding to Figure 3.8

Let an associated BIGS have $\mathcal{B}^* = (F, \Omega; H^*)$, where $\beta_\kappa^* = F_\kappa$, and let $H^*$ consist of the links from transactions to companies directly, such that $\Omega_s(\mathcal{B}) = \Omega_s$. Here the superscript is used to distinguish $\beta_\kappa^*$ in $\mathcal{B}^*$ from $\beta_\kappa$ in $G$. This gives rise to (I)* - (III)* in Figure 3.9 corresponding to (I) - (III) in Figure 3.8.

### 3.4.2  Estimation

The HTE (2.6) based on $\Omega_s$ is the same either calculated with respect to $G$ (as in Figure 3.8) or $\mathcal{B}^*$ (as in Figure 3.9).

One can apply another IWE under BIGS from $\mathcal{B}^*$. Due to the dependence on $\mathcal{B}^*$, the HH-type estimator (2.8) can be given as

$$\hat{\theta}_z^* = \sum_{i \in s_0} \frac{z_i^*}{\pi_i} \qquad \text{and} \qquad z_i^* = \sum_{\kappa \in \alpha_i^*} \omega_{i\kappa}^* y_\kappa \tag{3.3}$$

where $\sum_{i \in \beta_\kappa^*} \omega_{i\kappa}^* = 1$ for any $\kappa \in \Omega$, and $\alpha_i^*$ and $\beta_\kappa^*$ refer to $\mathcal{B}^*$.

There are other ways of using fixed incident weights defined for $\mathcal{B}_1$ and $\mathcal{B}_2$, respectively, where $\omega_{ib}$ for $i \in F$ and $b \in \alpha_i \subset Q$ and $\sum_{i \in \beta_b} \omega_{ib} = 1$ for each $b \in Q$, and $\omega_{b\kappa}$ for $b \in Q$ and $\kappa \in \alpha_b \subseteq \Omega$ and $\sum_{b \in \beta_\kappa} \omega_{b\kappa} = 1$ for each $\kappa \in \Omega$.

Let $\delta_b = 1$ if $b \in Q$ is sampled and 0 otherwise. Let the *HT-HH estimator* be given by

$$\hat{\theta}_{yz} = \sum_{b \in Q} \frac{\delta_b}{\pi(b)} z_b \qquad \text{and} \qquad z_b = \sum_{\kappa \in \alpha_b} \omega_{b\kappa} y_\kappa \tag{3.4}$$

That is, a value $z_b$ is constructed for $b \in Q$ using the HH-type weights between $Q$ and $\Omega$, which is then used in the manner of HTE based on the links between $F$ and $Q$. The estimator (3.4) is unbiased, since

$$E(\hat{\theta}_{yz}) = \sum_{b \in Q} z_b = \sum_{\kappa \in \Omega} y_\kappa$$

Let $\omega_{i\kappa} = \sum_{b \in \alpha_i} \omega_{ib}\omega_{b\kappa}$ for $i \in F$ and $\kappa \in \alpha_i^2$. Let the *HH-HH estimator* be given by

$$\hat{\theta}_{zz} = \sum_{i \in s_0} \frac{z_i}{\pi_i} \quad \text{and} \quad z_i = \sum_{\kappa \in \alpha_i^2} \omega_{i\kappa} y_\kappa = \sum_{b \in \alpha_i} \sum_{\kappa \in \alpha_b} \omega_{ib}\omega_{b\kappa} y_\kappa \quad (3.5)$$

That is, $y_\kappa$ is apportioned to $z_{b\kappa} = \omega_{b\kappa} y_\kappa$ for the intermediary nodes $b$, and $z_{b\kappa}$ is apportioned to $z_i$ by the weights $\omega_{ib}$, according to which $y_\kappa$ is apportioned to $z_i$ by the compound weights $\omega_{i\kappa}$. The estimator (3.5) is unbiased, since

$$E(\hat{\theta}_{zz}) = \sum_{i \in F} z_i = \sum_{i \in F} \sum_{b \in \alpha_i} \sum_{\kappa \in \alpha_b} \omega_{ib}\omega_{b\kappa} y_\kappa$$

$$= \sum_{\kappa \in \Omega} y_\kappa \sum_{b \in \beta_\kappa} \omega_{b\kappa} \sum_{i \in \beta_b} \omega_{ib} = \sum_{\kappa \in \Omega} y_\kappa$$

### 3.4.3 A special case

The estimators (3.3), (3.4) and (3.5) are different ways of applying the strategy BIGS-IWE in the current situation. In the special case of $|s_0| = 1$, they become identical if the PIDA weights (2.10) with $\gamma = 0$ is applied everywhere, yielding

$$\omega_{i\kappa}^* = \frac{\pi_i}{\sum_{j \in \beta_\kappa^*} \pi_j} \quad \omega_{ib} = \frac{\pi_i}{\sum_{j \in \beta_b} \pi_j} \quad \omega_{b\kappa} = \frac{\pi_{(b)}}{\sum_{c \in \beta_\kappa} \pi_{(c)}}$$

where

$$\pi_{(b)} = \sum_{j \in \beta_b} \pi_j \quad \text{and} \quad \sum_{c \in \beta_\kappa} \pi_{(c)} = \sum_{c \in \beta_\kappa} \sum_{j \in \beta_c} \pi_j$$

now that $|s_0| = 1$. Since $|\alpha_i| \equiv 1$, we have $\beta_c \cap \beta_b = \emptyset$ for $b \neq c \in Q$, such that

$$\sum_{c \in \beta_\kappa} \sum_{j \in \beta_c} \pi_j = \sum_{j \in \beta_\kappa^*} \pi_j$$

It follows that $w_{i\kappa}^* = \omega_{ib}\omega_{b\kappa}$ and $\hat{\theta}_z^* = \hat{\theta}_{zz}$. Since $a_{ib_i} = 1$ for only one $b_i \in Q$, we have

$$\hat{\theta}_{yz} = \sum_{i \in s_0} \sum_{b \in \alpha_i} \frac{z_{(b)}}{\pi_{(b)}} = \sum_{i \in s_0} \frac{1}{\pi_{(b_i)}} \sum_{\kappa \in \alpha_{b_i}} \frac{\pi_{(b_i)}}{\sum_{j \in \beta_\kappa^*} \pi_j} y_\kappa$$

$$= \sum_{i \in s_0} \sum_{\kappa \in \alpha_i^2} \frac{y_\kappa}{\sum_{j \in \beta_\kappa^*} \pi_j} = \sum_{i \in s_0} \sum_{\kappa \in \alpha_i^*} w_{i\kappa}^* \frac{y_\kappa}{\pi_i} = \hat{\theta}_z^*$$

To see that the choice matters, consider e.g. (III), where

$$\begin{cases} \omega_{b\kappa} = \omega_{b\ell} = \pi_h \\ \omega_{c\kappa} = \omega_{c\ell} = \pi_i + \pi_j \end{cases} \quad \begin{cases} \omega_{hb} = 1 \\ \omega_{ic} = \frac{\pi_i}{\pi_i + \pi_j} \\ \omega_{jc} = \frac{\pi_j}{\pi_i + \pi_j} \end{cases} \Rightarrow \begin{cases} z_b = 2\pi_h \\ z_c = 2(\pi_i + \pi_j) \end{cases}$$

such that setting $\gamma = 0$ in (2.10) yields

$$\hat{\theta}_{yz} = \frac{\delta_b}{\pi_h} 2\pi_h + \frac{\delta_c}{\pi_i + \pi_j} 2(\pi_i + \pi_j) \equiv 2 \quad \text{and} \quad V(\hat{\theta}_{yz}) = 0$$

Whereas $V(\hat{\theta}_{yz}) > 0$ using the multiplicity weights between $Q$ and $\Omega$, which are $\omega_{b\kappa} = \omega_{b\ell} = \omega_{c\kappa} = \omega_{c\ell} = 0.5$, since

$$z_b = z_c = 1 \quad \Rightarrow \quad \hat{\theta}_{yz} = \frac{\delta_b}{\pi_h} + \frac{\delta_c}{\pi_i + \pi_j}$$

## BIBLIOGRAPHIC NOTES

Zhang and Oguz-Alper (2020) establish a version of Theorem 3.1, which specifies the conditions for applying the strategy BIGS-IWE to arbitrary graph sampling situations. The formulation is revised here to avoid potential confusions based on feedbacks received in private communications.

The multiplicity estimator of Birnbaum and Sirken (1965) is common in indirect or network sampling. Frank (1977c) notes that such methods "are not explicitly stated as graph problems but which can be given such formulations".

For baseline-LIS, Becker (1991) applies the HH-estimator, and Thompson (2012, Ch. 19.1) describes the details of the HTE. As illustrated in Patone and Zhang (2020) and here in Section 3.3, both are special cases of the strategy BIGS-IWE to LIS and many other unbiased estimators can be considered. The approach can be extended to the general setting (Kaiser, 1983), where a point is randomly selected on the map, and an angle is randomly chosen, yielding a line in the chosen direction.

All these so-called unconventional sampling techniques are characterised by the presence of some deterministic observation rules *in addition* to the initial sampling design. The application of the strategy BIGS-IWE helps to clarify the common structure underlying these seemingly unrelated problems, consisting of the distinction between sampling units (household or line) and study units (sibling network or wild-life habitat), and the many-many incident observation links between them.

Given the ancestor set $F_\kappa$ for each study unit in the sample graph, one can freely fix $\beta_\kappa \subseteq F_\kappa$ with respect to efficiency. Notice that two examples are given in this chapter, however, where the knowledge of $F_\kappa$ is incomplete. For sampling of sibling networks via households, one can use $\beta_\kappa \subseteq F_{\kappa|s}$ if the ancestor households outside the initial sample are not identified. For baseline-LIS, one can use $\mathcal{B}^*$ constructed from the actual sample $\Omega_s$ to approximate the calculation using $\mathcal{B}'$ based on detectability partitions.

# CHAPTER 4

# Adaptive cluster sampling

Starting from an initial probability sample of units, *adaptive cluster sampling (ACS)* refers to sampling designs in which, whenever the value of interest of a selected unit satisfies a given criterion, other units in the "neighbourhood" of that unit are added to the sample.

*Adaptive network sampling (ANS)* is network sampling using an adaptive OP. Here we consider ANS which can be given as ACS from graphs, where the neighbourhood of a node consists of its adjacent nodes and each cluster is a network in the graph. The population graph may or may not be known, insofar as the values needed for the adaptive OP are unknown generally.

## 4.1   SPATIAL ACS

In the example of Thompson (1990), the population $U$ consists of 5 spatial grids, with associated $y_U = \{1, 0, 2, 10, 1000\}$ for the amount of species of interest. Each grid has one or two contiguous grids as its neighbours, which are adjacent in the undirected simple graph $G = (U, A)$ in Figure 4.1, where we simply denote each grid by its $y$-value and refer to it as a node. The graph $G$ is known, but the associated $y$-values are unknown.

Given an initial sample $s_0$ of size 2 by SRS from $F = U$, one would survey their $y$-values, and all the neighbours of a sample

$$G: \quad 1 \underline{\quad} 0 \underline{\quad} 2 \underline{\quad} 10 \underline{\quad} 1000$$

**Figure 4.1** A graph for spatial ACS (Thompson, 1990)

node $i$ only if $y_i \geq 5$. The OP is repeated for all the acquired nodes, which may or may not generate additional nodes to be surveyed. The process is terminated, when the last observed nodes are all below the threshold. The interest is to estimate the mean of $y$ over $U$, denoted by $\mu$. In terms of (1.1) or (1.2), we have

$$\Omega = U \quad \text{and} \quad \theta = \sum_{\kappa \in \Omega} y_\kappa \quad \text{and} \quad \mu = \theta/N$$

A network in such spatial graphs may consist of a single node, whose $y$-value is below the threshold, to be referred to as a *terminal node*, because ACS would not proceed further from such a node; or it may consist of a cluster of nodes, where $y_i \geq 5$ for every node $i$ in the network, to be referred to as a *non-terminal (NT) network*. An edge node is a terminal node that is adjacent to at least one NT network, such as 2 to $\{10, 1000\}$ in Figure 4.1, where 2 can be observed from 10 or 1000 but not *vice versa*.

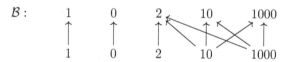

**Figure 4.2** Ancestors under ACS from Figure 4.1

ACS from Figure 4.1 is the same as BIGS from $\mathcal{B} = (F, \Omega; H)$ in Figure 4.2. While a non-edge terminal node has itself as the sole ancestor, such as node 1, the edge node 2 has itself and its adjacent NT-networks as the ancestors, $F_2 = \{2, 10, 1000\}$. The sample inclusion probability of a terminal node cannot be calculated without observing all its neighbours in $G$, lest any of them may belong to an NT network. The inclusion probability of a non-terminal node $\kappa$ is the probability that its NT-network (or ancestor set), such as $F_{10} = \{10, 1000\}$, intersects the initial sample.

Thompson (1990) defines a *modified HTE*, where a terminal node $\kappa$ is *eligible* for estimation *only if* it is selected in $s_0$ directly. This probability $\pi_\kappa = \Pr(\kappa \in s_0)$ is known. The modified HTE can

be given as

$$\hat{\theta}_{HT}^* = \sum_{\kappa \in \Omega_s} W_\kappa y_\kappa = \sum_{\kappa \in \Omega} \mathbb{I}(\kappa \in \Omega_s) W_\kappa y_\kappa$$

where $W_\kappa^{-1} = \Pr(\kappa \in \Omega_s)$ for any non-terminal node $\kappa$, and

$$W_\kappa = \begin{cases} \pi_\kappa^{-1} & \text{if } \kappa \in s_0 \\ 0 & \text{otherwise} \end{cases}$$

for any terminal node $\kappa$. This strategy, denoted by $(G, \hat{\theta}_{HT}^*)$, is unbiased because $E\big(\mathbb{I}(\kappa \in \Omega_s) W_\kappa\big) = 1$, $\forall \kappa \in \Omega$.

### 4.1.1  Two strategies BIGS-IWE

Consider first associated BIGS from $\mathcal{B}^* = (F, \Omega; H^*)$ in Figure 4.3, where the NT-network ancestors are removed for any edge node in (3.2), such that $\beta_2^* = \{2\} \subset F_2$. In other words, we use the ancestor set $\beta_\kappa^* = \{\kappa\}$ that is known *a priori* for a terminal node $\kappa$, and $\beta_\kappa^* = F_\kappa$ that is always observed for a non-terminal node $\kappa$ whenever $\kappa \in \Omega_s$. Notice that a terminal node $\kappa$ may be included in $\Omega_s$ but not in $\Omega_s(\mathcal{B}^*)$. One can use the HTE (2.6) with respect to BIGS from $\mathcal{B}^*$. Denote this strategy by $(\mathcal{B}^*, \hat{\theta}_y)$.

**Figure 4.3**  BIGS with known ancestor for node 2 in Figure 4.1

The two strategies $(G, \hat{\theta}_{HT}^*)$ and $(\mathcal{B}^*, \hat{\theta}_y)$ yield actually the same estimator, where the sample nodes in $\Omega_s$ which are eligible for the former strategy are simply $\Omega_s(\mathcal{B}^*)$. The difference is that applying the RB method to $\hat{\theta}_y$ under BIGS from $\mathcal{B}^*$ does not change it, whereas RB adjustment of $\hat{\theta}_{HT}^*$ is possible under ACS from $G$.

It is possible to apply other IWEs under BIGS from $\mathcal{B}^*$. Now that an NT-network is observed entirely under ACS if any of its nodes is observed, $|\alpha_i|$ required for the PIDA weight (2.10) is the same for all the nodes in the same network. Given initial SRS, the estimator $\hat{\theta}_{z\alpha\gamma}$ reduces to the multiplicity estimator $\hat{\theta}_{z\beta}$.

Consider next associated BIGS from $\mathcal{B}^{\dagger} = (F, \Omega; H^{\dagger})$ in Figure 4.4, which differs to $\mathcal{B}^*$ in that only the NT-network ancestors are retained for an edge node in (3.2), such that $\beta_2^{\dagger} = \{10, 1000\} \subset F_2$. One can apply the HTE (2.6), yielding the strategy $(\mathcal{B}^{\dagger}, \hat{\theta}_y)$.

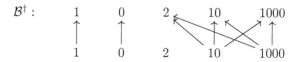

**Figure 4.4**   BIGS without known ancestor for 2 in Figure 4.1

Whereas the strategy $(\mathcal{B}^*, \hat{\theta}_y)$ can be determined *a priori* for ACS, the strategy $(\mathcal{B}^{\dagger}, \hat{\theta}_y)$ is *sample-dependent* in the sense that it can only be identified for the edge node 2 in Figure 4.1 when its NT-network $\{10, 1000\}$ intersects $s_0$, the probability of which is $1 - \bar{\pi}_{10,1000} = 0.7$. The probability is 0.2 that 2 is observed on its own while neither 10 nor 1000 is selected in $s_0$, in which case one needs to adopt $\mathcal{B}^*$ instead. In other words, for a sample-dependent BIGS, one can fix $\beta_\kappa$ in (3.2) for a terminal node $\kappa$ as follows:

- use $\beta_\kappa^* = \{\kappa\}$ if $\kappa$ is observed on its own,

- use an adjacent NT network $\beta_\kappa^{\dagger}$ if both $\kappa$ and $\beta_\kappa^{\dagger}$ are observed.

Given the node 2 is observed by ACS from Figure 4.1, the strategy $(\mathcal{B}^*, \hat{\theta}_y)$ is adopted 2 out 9 times, whereas $(\mathcal{B}^{\dagger}, \hat{\theta}_y)$ is adopted the rest of times. Since only one sample graph is realised in reality, inference needs to be conditional on the chosen strategy.

## 4.1.2   Discussion

For ACS from Figure 4.1, let $F_{\kappa|s}$ be the ancestors of a node $\kappa$, which are identified in the actual sample $G_s$. For an associated BIGS by (3.2), the choice $\beta_2^* = \{2\} \subseteq F_{2|s}$ can be determined *a priori*, whereas the choice $\beta_2^{\dagger}$ can only be identified if 2 is sampled together with $\beta_2^{\dagger} = \{10, 1000\}$. Indeed, $F_{2|s} = \{2, 10, 1000\}$ is identified together with $\beta_2^{\dagger}$, so one can also fix $\beta_2 = F_{2|s}$ in (3.2). Since $F_{2|s} = F_2$ here, the resulting $\hat{\theta}_y$ is actually the HTE under ACS from Figure 4.1, denoted by $(\mathcal{B}, \hat{\theta}_y)$.

Table 4.1 lists the details of the different strategies, including the observed sample under each strategy in addition to the initial sample $s_0$. In particular, the edge node 2 is given in italic in the 5 samples where it is eligible for $(\mathcal{B}, \hat{\theta}_y)$ but not $(G, \hat{\theta}^*_{HT})$.

The two estimators $\hat{\mu}^*_{HT}$ under $(G, \hat{\theta}^*_{HT})$ and $\hat{\mu}_y$ under $(\mathcal{B}^*, \hat{\theta}_y)$ differ only regarding the RB method. The difference hinges on the initial sample $s_0 = \{10, 1000\}$. Under $(G, \hat{\theta}^*_{HT})$, the same sample (including 2) is also observed from $s_0 = \{2, 10\}$ or $\{2, 1000\}$, but the estimate $\hat{\mu}^*_{HT}$ differs, because 2 is unused when $s_0 = \{10, 1000\}$. The RB method yields $\hat{\mu}^*_{HTRB} = 289.238$ given $\Omega_s = \{2, 10, 1000\}$. In contrast, under $(\mathcal{B}^*, \hat{\theta}_y)$ the estimate $\hat{\mu}_y$ is unchanged by the RB method, because the observed sample $\Omega_s(\mathcal{B}^*)$ from $s_0 = \{10, 1000\}$ differs to that from $s_0 = \{2, 10\}$ or $\{2, 1000\}$.

The BIGS inclusion probability of the edge node 2 is raised to 0.7 under $(\mathcal{B}^\dagger, \hat{\theta}_y)$; it is further raised to 0.9 under $(\mathcal{B}, \hat{\theta}_y)$. Neither is a good choice for ACS, because the value of an edge node is below the threshold by definition, and the resulting sampling variances are somewhat larger than by $(\mathcal{B}^*, \hat{\theta}_y)$.

### 4.1.3 Two-stage ACS

The left plot in Figure 4.5 gives the setting of two-stage ACS in Thompson (1991). Each vertical strip is a primary sampling unit, and each grid a secondary sampling unit. Given a strip selected at the first stage, all the grids belonging to it are searched for the species. Next, neighbouring grids to those with species are searched, and so on. Thus, ACS is applied at the second stage, which is terminated once no more non-empty grids (with species) are found in this way. An edge grid is an empty grid (without species), which is contiguous to one or more non-empty grids.

The right plot in Figure 4.5 gives the associated BIGS, similarly defined as $\mathcal{B}^*$ in Figure 4.3, where $F$ consists of the strips and $\Omega$ the grids. Each big node marked by a capital letter denotes a strip. The small nodes denote the grids. There are 10 star-like subgraphs, where a strip is adjacent to its 20 empty grids, which are observed under BIGS only if this strip is selected in $s_0$. The small nodes (darker in shade) that are adjacent to four big nodes form a cluster of non-empty grids, which are all observed if any of the four strips are selected in $s_0$. There are three such clusters of non-empty grids. Each of the 10 strips that contains non-empty grids is also adjacent

Table 4.1 Strategies for ACS from $G$ in Figure 4.1.

| $s_0$ | $(G, \hat{\theta}^*_{HT})$ or $(\mathcal{B}, \hat{\theta}_y)$ | | | $(\mathcal{B}^*, \hat{\theta}_y)$ | | $(\mathcal{B}^\dagger, \hat{\theta}_y)$ | |
|---|---|---|---|---|---|---|---|
| | $\Omega_s = \Omega_s(\mathcal{B})$ | $\hat{\mu}^*_{HT}$ | $\hat{\mu}_y$ | $\Omega_s(\mathcal{B}^*)$ | $\hat{\mu}_y$ | $\Omega_s(\mathcal{B}^\dagger)$ | $\hat{\mu}_y$ |
| 1,0 | 1,0 | 0.500 | 0.500 | 1,0 | 0.500 | 1,0 | 0.500 |
| 1,2 | 1,2 | 1.500 | 0.944 | 1,2 | 1.500 | 1 | 0.500 |
| 0,2 | 0,2 | 1.000 | 0.444 | 0,2 | 1.000 | 0 | 0.000 |
| 1,10 | 1,10,2,1000 | 289.071 | 289.516 | 1,10,1000 | 289.071 | 1,10,2,1000 | 289.643 |
| 1,1000 | 1,1000,2,10 | 289.071 | 289.516 | 1,1000,10 | 289.071 | 1,1000,2,10 | 289.643 |
| 0,10 | 0,10,2,1000 | 288.571 | 289.016 | 0,10,1000 | 288.571 | 0,10,2,1000 | 289.143 |
| 0,1000 | 0,1000,2,10 | 288.571 | 289.016 | 0,1000,10 | 288.571 | 0,1000,2,10 | 289.143 |
| 2,10 | 2,10,1000 | 289.571 | 289.016 | 2,10,1000 | 289.571 | 2,10,1000 | 289.143 |
| 2,1000 | 2,1000,10 | 289.571 | 289.016 | 2,1000,10 | 289.571 | 2,1000,10 | 289.143 |
| 10,1000 | 10,1000,2 | 288.571 | 289.016 | 10,1000 | 288.571 | 10,1000,2 | 289.143 |
| Variance | 17418.4 | | 17482.4 | 17418.4 | | 17533.7 | |

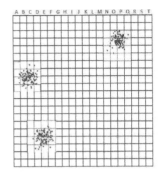 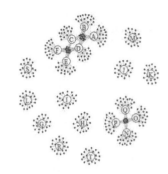

**Figure 4.5**  Two-stage ACS (Thompson, 1991), left; BIGS, right

to the rest of its empty grids (lighter in shade). An edge grid is not adjacent to its neighbour strip in $\mathcal{B}^*$, although it can be observed via the latter under two-stage ACS.

Notice that the choice of $\beta_\kappa$ in (3.2) does not matter for any edge node $\kappa$ in this case, because $y_\kappa = 0$ does not contribute to the total estimation regardless the inclusion probability of $\kappa$.

The total of interest in terms of (1.2) is given by

$$\theta = \sum_{\kappa \in \Omega} y_\kappa = 326$$

The initial sample $s_0$ of strips are obtained by SRS, where $m = |s_0|$. Consider the estimators below:

- the HTE $\hat{\theta}$ only based on the $m$ stripes directly selected in $s_0$;

- the HTE $\hat{\theta}_y$ by (2.6) under BIGS from $\mathcal{B}^*$, where $|F| = 20$ and $\pi_{(\kappa)} = 1 - \binom{16}{m} / \binom{20}{m}$ for a non-empty grid $\kappa$;

- the multiplicity estimator $\hat{\theta}_{z\beta}$ with the weights $\omega_{i\kappa} = 1/4$;

- the HH-type estimator $\hat{\theta}_{z\alpha}$ using PIDA-like weights, where $a_{1+}^{-\gamma}$ is replaced by $1/2$ for strips C and D, and 1 for the other non-empty strips. These PIDA-like weights can be set without the additional effort that is required to obtain $\{|\alpha_i| : i \in s_0\}$.

The standard errors are presented in Table 4.2 for $m = 1, ..., 10$. The estimator $\hat{\theta}$ without ACS at the 2nd-stage has the largest standard error given any $m$. The HTE $\hat{\theta}_y$ becomes more efficient

Table 4.2   Standard errors of selected estimators under two-stage
ACS given initial SRS with $m = |s_0|$

| $m$ | $\hat{\theta}$ | $\hat{\theta}_y$ | $\hat{\theta}_{z\beta}$ | $\hat{\theta}_{z\alpha}$ |
|-----|------|------|------|------|
| 1   | 457  | 356  | 356  | 329  |
| 2   | 315  | 236  | 245  | 226  |
| 3   | 250  | 179  | 194  | 180  |
| 4   | 210  | 142  | 163  | 151  |
| 5   | 182  | 116  | 141  | 131  |
| 6   | 160  | 95   | 125  | 115  |
| 7   | 143  | 78   | 111  | 103  |
| 8   | 128  | 63   | 100  | 92   |
| 9   | 116  | 51   | 90   | 83   |
| 10  | 105  | 40   | 82   | 75   |

than both $\hat{\theta}_{z\beta}$ and $\hat{\theta}_{z\alpha}$ as $m$ increases. The estimator $\hat{\theta}_{z\alpha}$ is the
most efficient if $m < 3$, and it is always more efficient than the
multiplicity estimator $\hat{\theta}_{z\beta}$. There are infinite ways of constructing
the HH-type estimator, some of which may well be more efficient
than the estimators illustrated here.

## 4.2   EPIDEMIC PREVALENCE ESTIMATION

Consider a setting of epidemiological study from $G = (U, A)$ where
only $U$ is known but $A$ and the associated values $\{y_i : i \in U\}$ are
unknown. Provided the virus is transmitted via personal contacts,
let $(ij), (ji) \in A$ if relevant contact exists between individuals $i$
and $j$, such that $A$ contains all the relevant contacts in $U \times U$.
The population graph $G = (U, A)$ is undirected and simple. Let
$N = |U|$ and $F = U$. Let $y_i = 1$ if person $i \in U$ is a *case*, and
$y_i = 0$ otherwise. Let the population *prevalence* be

$$\mu = \theta/N \qquad \text{where} \qquad \theta = \sum_{i \in U} y_i$$

### 4.2.1   ACS from $G$ with unknown $A$

The cases are clustered by the contacts in $A$. ACS can increase
the sample yield of cases with $y_i = 1$ relatively to *noncases* with
$y_i = 0$, in order to improve the design efficiency. Let $s_0$ be an initial
sample from $U$, with $\pi_i = \Pr(i \in s_0)$. By *adaptive network tracing*,

all the contacts of each case $i$ in $s_0$ are included, and the procedure is repeated for them, and so on until no more cases can be added in this way. Let $\pi_{(i)}$ be the inclusion probability in the final (seed) sample. Since any case $i$ that is in contact with other cases can be sampled by adaptive network tracing, even when it is not selected in $s_0$ directly, we achieve $\pi_{(i)} > \pi_i$.

The sample $s$ contains cases and noncases. The cases can be partitioned into *case networks*, where the individuals in the same case network all have $y_i = 1$ and are connected either directly or via other cases. For any case $i$, the ACS sample inclusion probability $\pi_{(i)}$ is the probability that the intersection of $s_0$ and its network is non-empty. Each noncase with $y_i = 0$ may be directly selected in $s_0$ or via its adjacent case network. Its inclusion probability $\pi_{(i)}$ cannot be calculated generally. However, this does not matter, since a noncase contributes 0 to the estimation of $\theta$ regardless.

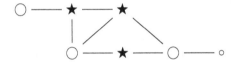

**Figure 4.6** Illustration of case (★) edge node (○) and other non-case (○) in $G$

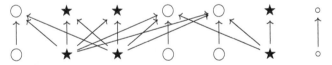

**Figure 4.7** BIGS corresponding to Figure 4.6

Figure 4.6 illustrates two networks of cases (★), their edge nodes (○) and another noncase (○) in $G$. ACS from Figure 4.6 is the same as BIGS from Figure 4.7. In particular, any ★ would lead to its adjacent ○ but not *vice versa*. Figure 4.8 shows the associated BIGS, where $\Omega$ consists only of the case networks, after removing the noncases.

In terms of BIGS as illustrated in Figure 4.8, let $\kappa$ denote the network of case $i$, with nodes $\beta_\kappa$ and $n_\kappa = |\beta_\kappa|$. The HTE of $\theta$ can be given by

$$\hat{\theta}_y = \sum_{i \in s} \frac{y_i}{\pi_{(i)}} = \sum_{\kappa \in \Omega_s} \frac{n_\kappa}{\pi_{(\kappa)}} \qquad (4.1)$$

**Figure 4.8** BIGS with case networks as study units for ACS from Figure 4.6

where $\pi_{(\kappa)} = \pi_{(i)}$ for any $i \in \beta_\kappa$. We have

$$V(\hat{\theta}_y) = \sum_{\substack{i \in U \\ y_i = 1}} \sum_{\substack{j \in U \\ y_j = 1}} \left( \frac{\pi_{(ij)}}{\pi_{(i)}\pi_{(j)}} - 1 \right)$$

$$= \sum_{\kappa \in \Omega} \sum_{\ell \in \Omega} \left( \frac{\pi_{(\kappa\ell)}}{\pi_{(\kappa)}\pi_{(\ell)}} - 1 \right) n_\kappa n_\ell \qquad (4.2)$$

since only $y_i y_j = 1$ contribute to the double summation.

For the HH-type estimator, we have $|\alpha_i| = n_\kappa$ for any $i \in \beta_\kappa$. Provided $\pi_{(\kappa)} \approx \sum_{i \in \beta_\kappa} \pi_i$ and $\pi_i \approx \pi_j$ for $i \neq j \in \beta_\kappa$, there would be little difference to the HTE.

The network-exhaustive OP of ACS can be an issue if a network is too large to be surveyed completely due to practical reasons. If it is possible to measure $\xi_{ij}$ as the *strength* of $(ij) \in A$, then one may define *adaptively* the neighbourhood of $i$ to be

$$\nu_i^* = \{j \in U : (ij) \in A, \xi_{ij} > \xi_0\}$$

for a chosen threshold $\xi_0$, and include $\nu_i^*$ if $y_i = 1$. The resulting sampling method may be referred to as *doubly adaptive cluster sampling*, since it is based on two threshold values $y_i > 0$ and $\xi_{ij} > \xi_0$. Imposing a maximum number of waves, say $q$, is another way to curtail large networks. The sampling is terminated after the $q$-th wave, even if $s_q \neq \emptyset$, yielding *q-wave adaptive snowball sampling*. As a result of either modification, a case with $y_i = 1$ may become a terminal node that no longer can be ignored as when a terminal node always has $y_i = 0$. Strategies that can handle such problems will be described in Chapter 5.

## 4.2.2 Some simulation results

As illustrated below, one may use simulations to study the salient design aspects, for which purpose we shall concentrate on the HTE.

### 4.2.2.1  Size-biased sampling and adaptive network tracing

Let $\eta$ be the odds of case selection in the initial sample $s_0$, which is defined as the ratio of the probability that a case is included in $s_0$ against that of a noncase. Initial sampling is *size-biased* if $\eta \neq 1$, positively so if $\eta > 1$.

Table 4.3 presents some results for the RE of ACS, defined as the ratio of the variance of the HTE under ACS against that based on the initial sample $s_0$, which is either selected by SRS ($\eta = 1$) or sized-biased sampling with $\eta = 2$. All the cases in the population are divided into networks, which have the same size $c$.

- ACS is increasingly more efficient than SRS as $m$ increases, if one compares the CVs of SRS with $m$ and ACS with $E(n)$, where $m \approx E(n)$. The gain is more pronounced given large case networks. Given initial SRS of size 10000, ACS requires about 900 extra individuals, by which the sampling variance is reduced to 0.00, because any case network almost certainly intersects $s_0$. The reduction is quicker given initial size-biased sampling, where the variance is already 0.00 at $m = 5000$.

- ACS has basically no gains given only small case networks with $c = 2$, where size-biased sampling would be the chief means for reducing variance. Given $\eta = 2$ the variance of the initial sample estimator is about halved given any $m$ in Table 4.3.

In summary, positive size-biased sampling and adaptive network tracing can enhance each other, generating extra gains of design efficiency when they are applied jointly.

The inclusion probabilities are easy to compute under initial SRS. For unequal probability sampling of $s_0$, where $\bar{\pi}_{\beta_\kappa}$ is unknown if $|\beta_\kappa| > 1$, it is convenient to treat the initial sampling as if it were Poisson sampling. It has been verified empirically that the approximation holds well in the simulation settings here, including the highest sampling fraction 10%.

### 4.2.2.2  Population of households

Contact is hard to avoid within a household. Denote by $\mathbb{G} = (\mathbb{H}, \mathbb{A})$ the population graph, where $\mathbb{H}$ consists of all the households, and $\mathbb{A}$ the contacts between any two households via their members. Size-biased sampling and adaptive network tracing in $\mathbb{G}$ follow the

Table 4.3 ACS given equal-size ($c$) case networks in population of size $N = 10^5$, $\theta = 10^3$ and prevalence $\mu = 0.01$. Initial sample of size $m$ by SRS ($\eta = 1$) or size-biased sampling ($\eta = 2$), ACS with sample size $n = |s|$ by adaptive network tracing

| SRS ($\eta=1$) | | ACS, $c = 100$ | | | ACS, $c = 10$ | | | ACS, $c = 2$ | | |
|---|---|---|---|---|---|---|---|---|---|---|
| $m$ | CV | $E(n)$ | CV | RE | $E(n)$ | CV | RE | $E(n)$ | CV | RE |
| 1000 | 0.31 | 1631 | 0.24 | 0.58 | 1085 | 0.31 | 0.96 | 1010 | 0.31 | 0.99 |
| 1630 | 0.24 | 2423 | 0.15 | 0.40 | 1766 | 0.24 | 0.93 | 1646 | 0.24 | 0.99 |
| 2420 | 0.20 | 3306 | 0.10 | 0.23 | 2614 | 0.19 | 0.89 | 2443 | 0.20 | 0.99 |
| 5000 | 0.14 | 5944 | 0.02 | 0.03 | 5352 | 0.12 | 0.79 | 5048 | 0.14 | 0.97 |
| 10000 | 0.09 | 10900 | 0.00 | 0.00 | 10551 | 0.07 | 0.59 | 10090 | 0.09 | 0.95 |

| Size-biased ($\eta=2$) | | ACS, $c = 100$ | | | ACS, $c = 10$ | | | ACS, $c = 2$ | | |
|---|---|---|---|---|---|---|---|---|---|---|
| $m$ | CV | $E(n)$ | CV | RE | $E(n)$ | CV | RE | $E(n)$ | CV | RE |
| 1000 | 0.22 | 1840 | 0.13 | 0.32 | 1160 | 0.21 | 0.91 | 1020 | 0.22 | 0.99 |
| 5000 | 0.09 | 5901 | 0.00 | 0.00 | 5549 | 0.07 | 0.60 | 5090 | 0.09 | 0.95 |
| 10000 | 0.06 | 10802 | 0.00 | 0.00 | 10692 | 0.04 | 0.31 | 10159 | 0.06 | 0.89 |

**Table 4.4** ACS given equal-size (c) case networks in household population of size $N = 10^5$, $\theta = 10^3$ and prevalence $\mu = 0.01$. Initial sample of size m by SRS ($\eta = 1$) or size-biased sampling ($\eta = 2$), ACS with sample size $n = |s|$ by adaptive network tracing

| SRS ($\eta = 1$) | | ACS, $c = 100$ | | | ACS, $c = 10$ | | | ACS, $c = 2$ | | |
|---|---|---|---|---|---|---|---|---|---|---|
| m | CV | $E(n)$ | CV | RE | $E(n)$ | CV | RE | $E(n)$ | CV | RE |
| 1000 | 0.35 | 1628 | 0.24 | 0.45 | 1086 | 0.31 | 0.77 | 1010 | 0.34 | 0.89 |
| 5000 | 0.16 | 5944 | 0.02 | 0.03 | 5351 | 0.13 | 0.63 | 5048 | 0.14 | 0.87 |
| 10000 | 0.11 | 10900 | 0.00 | 0.00 | 10552 | 0.07 | 0.49 | 10090 | 0.10 | 0.84 |

| Size-biased ($\eta = 2$) | | ACS, $c = 100$ | | | ACS, $c = 10$ | | | ACS, $c = 2$ | | |
|---|---|---|---|---|---|---|---|---|---|---|
| m | CV | $E(n)$ | CV | RE | $E(n)$ | CV | RE | $E(n)$ | CV | RE |
| 1000 | 0.25 | 1844 | 0.13 | 0.26 | 1162 | 0.22 | 0.71 | 1019 | 0.23 | 0.86 |
| 5000 | 0.11 | 5901 | 0.00 | 0.00 | 5547 | 0.08 | 0.47 | 5089 | 0.10 | 0.85 |
| 10000 | 0.07 | 10802 | 0.00 | 0.00 | 10691 | 0.04 | 0.25 | 10159 | 0.06 | 0.80 |

same definition as in $G = (U, A)$, but the actual design effects will differ between the two set-ups.

Table 4.4 presents some results for the RE of ACS. The only difference to Table 4.3 is that sampling and network tracing are from a population of households instead of persons. The stipulated household size distribution is $(0.38, 0.30, 0.12, 0.20)$ for household size $(1, 2, 3, 4)$, the same among both case and noncase households. The differences of $E(n)$ in Table 4.3 and 4.4 reflect the magnitude of simulation error in these results.

The variances are larger compared to sampling of individuals. However, the increases are smaller under ACS, such that the RE gains by adaptive network tracing are actually increased compared to sampling of persons, where the RE is appreciable even when the networks are small, e.g. $c = 2$. The RE of size-biased initial sampling is similar to that in Table 4.3.

## 4.3 ACS DESIGNS OVER TIME

Denote by $G_t = (U_t, A_t)$ the population graph for time point $t = 1, 2, \ldots$ in epidemiological studies. Even when $U_t$ is fixed, the contacts $A_t$ that are relevant for the time point $t$ will be dynamic, unless a strict lockdown is maintained over time.

### 4.3.1 Some basic designs

There are more than one possible definition of $U_t$, three of which are as given below.

a. Let $U_t = U$ be fixed over time.

b. Let $U_t$ be the union of $U_F$ and those that can be linked to $U_F$ via $A_t$, where $U_F$ is a fixed known population.

c. Let $U_t$ be the union of $U_{t-1}$ and those that can be linked to it via $A_t$ recursively, and $U_1$ is either by (a) or (b).

*Panel* The sample once selected is fixed over time, which only accommodates the definition (a) above. When the change $\mu_2 - \mu_1$ is estimated based on two independent samples at $t = 1, 2$, the variance is the sum of those of $\hat{\mu}_1$ and $\hat{\mu}_2$. The variance is reduced, when the panel design induces a positive correlation between $\hat{\mu}_1$ and $\hat{\mu}_2$. Suppose that based on two independent samples of the same size $n$, the sampling variance of the change estimator is

$$V(\hat{\mu}_2 - \hat{\mu}_1) = \frac{1}{n}\left(\mu_2(1 - \mu_2) + \mu_1(1 - \mu_1)\right) \doteq \frac{1}{n}\left(\mu_2 + \mu_1\right)$$

as long as $1 - \mu_2 \doteq 1$ and $1 - \mu_1 \doteq 1$. Let $\lambda_+$ be the proportion of the new cases over time, and $\lambda_-$ that of the closed cases, where $\lambda_+ \leq \mu_2$ and $\lambda_- \leq \mu_1$. Let $e_i = 0$ if individual $i$ has no change of case status, $e_i = 1$ if $i$ becomes a case, and $e_i = -1$ if $i$ becomes a closed case, such that $\mu_2 - \mu_1$ is the population mean of $e_i$. Suppose $\mu_2 = \mu_1$ for simplicity. It can be shown that

$$V_{panel}(\hat{\mu}_2 - \hat{\mu}_1) \doteq \frac{1}{n}(\lambda_+ + \lambda_-) \leq \frac{1}{n}\left(\mu_2 + \mu_1\right) = V(\hat{\mu}_2 - \hat{\mu}_1)$$

where $V_{panel}(\hat{\mu}_2 - \hat{\mu}_1)$ is the variance based on a panel of size $n$.

*Panel ACS (pACS)*   Only the initial sample $s_0$ is fixed over time, but the sample $s(t)$ obtained based on $s_0$ and $A_t$ via adaptive network tracing varies with $t$. This design is applicable to either the target population definition (a) or (b) above.

*Iterated ACS (iACS)*   Let $s(t)$ be the sample by ACS based on $s_0$ and $A_t$ for time $t$, with inclusion probabilities $\pi_i(t)$ and $\pi_{ij}(t)$ for any $i, j \in s(t) \subset U_t$, and let $s(t + 1)$ be given by ACS based on $A_{t+1}$ for time $t + 1$ which starts from

$$s_{0t} = s_0 \cup \{\, i \in s(t) \setminus s_0 : y_{i,t} = 1\}$$

Design-based inference of $\mu_{t+1} - \mu_t$ is possible using $\pi_i(t)$ and $\pi_{ij}(t)$, although one cannot calculate the inclusion probabilities in $s(t + 1)$. This can accommodate all the definitions (a) - (c) above.

Figure 4.9 illustrates a part of graphs with 7 nodes over time $t = 1, 2$. Suppose only $2 \in s_0$, which is a case at $t = 1$ and becomes a closed case at $t = 2$. The different samples are given below.

- Panel: $\bigstar_2 \in s(1)$, $\bigcirc_2 \in s(2)$

- pACS: $\{\bigcirc_1, \bigstar_2, \bigstar_3, \bigcirc_4, \bigcirc_6\} \subset s(1)$, $\bigcirc_2 \in s(2)$

- iACS: $\{\bigcirc_1, \bigstar_2, \bigstar_3, \bigcirc_4, \bigcirc_6\} \subset s(1)$, $\{\bigcirc_2, \bigstar_3, \bigcirc_4, \bigcirc_5\} \subset s(2)$

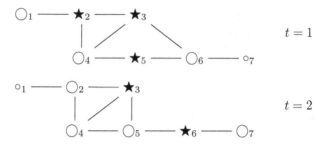

**Figure 4.9** Illustration of 7 nodes over time $t = 1, 2$

## 4.3.2 Estimation of change

The HTE of change $\nabla_{t,t+1} = \mu_{t+1} - \mu_t$ under the panel design is given by

$$\hat{\nabla}_{t,t+1}^{panel} = \frac{1}{N} \sum_{i \in s_0} \frac{1}{\pi_i}(y_{i,t+1} - y_{i,t})$$

where $\pi_i$ is the inclusion probability of $i \in s_0$ and $s_0$ is the panel.

The HTE of $\nabla_{t,t+1}$ under the panel ACS design is given by

$$\hat{\nabla}_{t,t+1}^{pACS} = \frac{1}{N_{t+1}} \sum_{i \in s(t+1)} \frac{y_{i,t+1}}{\pi_i(t+1)} - \frac{1}{N_t} \sum_{i \in s(t)} \frac{y_{i,t}}{\pi_i(t)}$$

where $s(t)$ is the sample at time $t$ by ACS based on $s_0$ and $A_t$, and $\pi_i(t)$ is the inclusion probability of $i \in s(t)$, and similarly for $s(t+1)$ and $\pi_i(t+1)$. That is, one applies (4.1) at each time point and takes the difference between them. The variance of each HTE follows from (4.2). For the covariance between them, we have

$$Cov(\hat{\mu}_t, \hat{\mu}_{t+1}) = \sum_{\substack{i \in U_t \\ y_{i,t}=1}} \sum_{\substack{j \in U_{t+1} \\ y_{j,t+1}=1}} \left( \frac{\pi_{ij}(t, t+1)}{\pi_i(t)\pi_j(t+1)} - 1 \right) \frac{y_{i,t}y_{j,t+1}}{N_t N_{t+1}}$$

where $\pi_{ij}(t, t+1) = \Pr\left(i \in s(t), j \in s(t+1)\right)$ refers to two population graphs, similarly as detailed below for the iACS design.

Under the iACS design, an unbiased estimator of $\nabla_{t,t+1}$ is

$$\hat{\nabla}_{t,t+1}^{iACS} = \frac{1}{N_{t+1}} \left( \sum_{\substack{i \in s_{0t} \\ y_{i,t}=1}} \frac{y_{i,t+1}}{\pi_i(t)} + \sum_{\substack{i \in s_{0t} \\ y_{i,t}=0}} \frac{y_{i,t+1}}{\pi_i} \right) - \frac{1}{N_t} \sum_{i \in s(t)} \frac{y_{i,t}}{\pi_i(t)}$$

The two terms in the parentheses form an HH-type estimator of

$\theta_{t+1}$, where the values $\{y_{j,t+1} : j \in s(t+1)\}$ are transformed to $\{z_i : i \in s_{0t}\}$. We have $z_i = 1$ using the multiplicity weights if $y_{i,t+1} = 1$ and $0$ if $y_{i,t+1} = 0$, so that $z_i = y_{i,t+1}$. Meanwhile, the inclusion probability in $s_{0t}$ differs whether an individual is case or not at $t$, since $s_{0t}$ is obtained by ACS based on $s_0$ and $A_t$, giving rise to the two terms above for emphasis.

The variance follows by considering the estimator $\hat{\nabla}^{iACS}_{t,t+1}$ as the HTE based on $s_{0t}$, with value $y_{i,t+1}/N_{t+1} - y_{i,t}/N_t$ for each $i \in s_{0t}$. Let $\kappa$ be the network of individual $i$, where $\beta_\kappa = \{i\}$ if $y_{i,t} = 0$. Let $\ell$ and $\beta_\ell$ for $j$ similarly. The joint inclusion probability of $i \neq j \in s(t)$ is given by

$$
\pi_{ij}(t) = \begin{cases}
\pi_{ij} & \text{if } y_{i,t} = 0, \ y_{j,t} = 0 \\
\pi_i + \pi_j(t) + \bar{\pi}_{\{i\} \cup \beta_\ell} - 1 & \text{if } y_{i,t} = 0, \ y_{j,t} = 1 \\
\pi_i(t) + \pi_j + \bar{\pi}_{\beta_\kappa \cup \{j\}} - 1 & \text{if } y_{i,t} = 1, \ y_{j,t} = 0 \\
\pi_i(t) + \pi_j(t) + \bar{\pi}_{\beta_\kappa \cup \beta_\ell} - 1 & \text{if } y_{i,t} = 1, \ y_{j,t} = 1
\end{cases}
$$

The estimator $\hat{\nabla}^{pACS}_{t,t+1}$ by panel ACS can be more efficient than $\hat{\nabla}^{panel}_{t,t+1}$ under the panel design, because $s_0$ is a subsample of either $s(t)$ or $s(t+1)$, and ACS increases the sample inclusion probability of a case. Likewise between $\hat{\nabla}^{iACS}_{t,t+1}$ and $\hat{\nabla}^{panel}_{t,t+1}$. The RE between panel and iterated ACS is undetermined in general. On the one hand, the sample $s(t+1)$ based on $s_0$ and $A_{t+1}$ is a subsample of that based on $s_{0t}$ and $A_{t+1}$, because, $s_0 \subseteq s_{0t} \subseteq s(t)$; on the other hand, the HH-type estimator of $\hat{\mu}_{t+1}$ under iterated ACS may be less efficient than the HTE under panel ACS. The strengths of the contrasting forces depend on how the case networks in $A_t$ and $A_{t+1}$ relate to each other.

### 4.3.3 Simulation results over two time points

New case networks may emerge from one time point to the next, whilst the existing ones may increase or decrease in their sizes. The speed may be quick or slow, at which a new case network emerges or an existing one grows or shrinks. Some settings over two time points are given in Table 4.5, where both the population size and prevalence are constant, such that the target parameter is $\nabla_{1,2} = 0$ in all the settings. The networks are all of size 2 at $t = 1$ in the last three settings S1 - S3: for those that are not growing, their sizes at

**Table 4.5** Population of constant size $N = 10^5$ and total $\theta = 10^3$. At $t = 1$, $\bar{\theta}$ case networks, each of equal size $c$; at $t = 2$, $(\bar{\theta}_+, c_+)$ or $(\bar{\theta}_-, c_-)$ existing networks of increasing or decreasing sizes, and $(\bar{\theta}', c')$ emerging networks

| Setting | Characterisation | $t = 1$ | $t = 2$ | | |
| --- | --- | --- | --- | --- | --- |
| | | $(\bar{\theta}, c)$ | $(\bar{\theta}_+, c_+)$ | $(\bar{\theta}_-, c_-)$ | $(\bar{\theta}', c')$ |
| L1 | Large, Quickly Evolving | (10, 100) | (2, 180) | (8, 80) | (0, 0) |
| L2 | Large, Quickly Emerging | (10, 100) | (0, 0) | (10, 80) | (2, 100) |
| L3 | Large, Slowly Emerging | (10, 100) | (0, 0) | (10, 90) | (5, 20) |
| M1 | Medium, Quickly Evolving | (100, 10) | (10, 46) | (90, 6) | (0, 0) |
| M2 | Medium, Quickly Emerging | (100, 10) | (0, 0) | (100, 6) | (10, 40) |
| M3 | Medium, Slowly Emerging | (100, 10) | (0, 0) | (100, 9) | (10, 10) |
| S1 | Small, Quickly Evolving | (500, 2) | (10, 42) | ($\leq 490, \leq 2$) | (0, 0) |
| S2 | Small, Quickly Emerging | (500, 2) | (0, 0) | ($\leq 500, \leq 2$) | (10, 40) |
| S3 | Small, Slowly Emerging | (500, 2) | (0, 0) | ($\leq 500, \leq 2$) | (50, 2) |

Table 4.6 Panel, panel ACS and iterated ACS designs for settings in Table 4.5

| (SE in $10^{-2}$) | Initial SRS, $m = 1000$ | | | | | | | | |
|---|---|---|---|---|---|---|---|---|---|
| | L1 | L2 | L3 | M1 | M2 | M3 | S1 | S2 | S3 |
| SE($\hat{\nabla}^{panel}_{t,t+1}$) | 0.20 | 0.20 | 0.14 | 0.28 | 0.28 | 0.14 | 0.28 | 0.28 | 0.13 |
| RE($\hat{\nabla}^{pACS}_{t,t+1}$) | 0.71 | 0.73 | 0.90 | 0.89 | 0.89 | 0.98 | 0.89 | 0.91 | 0.98 |
| RE($\hat{\nabla}^{iACS}_{t,t+1}$) | 0.57 | 0.60 | 0.52 | 0.69 | 0.70 | 0.52 | 0.84 | 0.85 | 0.75 |
| (SE in $10^{-2}$) | Initial SRS, $m = 5000$ | | | | | | | | |
| | L1 | L2 | L3 | M1 | M2 | M3 | S1 | S2 | S3 |
| SE($\hat{\nabla}^{panel}_{t,t+1}$) | 0.09 | 0.09 | 0.06 | 0.12 | 0.12 | 0.06 | 0.12 | 0.12 | 0.06 |
| RE($\hat{\nabla}^{pACS}_{t,t+1}$) | 0.09 | 0.11 | 0.38 | 0.62 | 0.63 | 0.87 | 0.65 | 0.64 | 0.91 |
| RE($\hat{\nabla}^{iACS}_{t,t+1}$) | 0.49 | 0.51 | 0.49 | 0.67 | 0.67 | 0.53 | 0.85 | 0.84 | 0.76 |
| (SE in $10^{-2}$) | Initial size-biased sampling ($\eta = 2$), $m = 1000$ | | | | | | | | |
| | L1 | L2 | L3 | M1 | M2 | M3 | S1 | S2 | S3 |
| SE($\hat{\nabla}^{panel}_{t,t+1}$) | 0.17 | 0.17 | 0.12 | 0.24 | 0.24 | 0.12 | 0.24 | 0.24 | 0.12 |
| RE($\hat{\nabla}^{pACS}_{t,t+1}$) | 0.31 | 0.33 | 0.41 | 0.51 | 0.51 | 0.50 | 0.52 | 0.53 | 0.51 |
| RE($\hat{\nabla}^{iACS}_{t,t+1}$) | 0.70 | 0.69 | 0.68 | 0.79 | 0.81 | 0.68 | 0.89 | 0.90 | 0.84 |

$t = 2$ are randomly assigned, subjected to the case total $\theta_2 = 10^3$, such that some of them may simply disappear by chance.

Table 4.6 presents some simulation results for the settings in Table 4.5, where the RE of an adaptive design is calculated against the panel design.

- Overall, from top-right towards bottom-left in Table 4.6, the RE of panel ACS improves quickly with the three initial values $(c, m, \eta)$. The RE of iterated ACS improves with $m$ except for S1-S3, but not with $\eta$, although it remains more efficient than the panel design as $\eta$ increases. The improvements are larger for panel ACS than iterated ACS.

- For any given initial network size $c$, moving between the three patterns of case networks over time, the RE of pACS improves less for slowly than quickly changing networks, as $m$ increases. As the initial odds of case selection $\eta$ increases, the RE of the panel ACS become more uniform across all the patterns.

- Given any combination of initial values of $(c, m, \eta)$, the RE of iterated ACS becomes more uniform across the three patterns of case networks, as $m$ and $\eta$ increase.

- Between the two ACS designs, the panel ACS design is more efficient given sufficiently large initial sample size $m$ and high odds of case selection $\eta$, whereas the iterated ACS design is more efficient given small $m$ and initial SRS, especially given slowly changing networks, where the panel ACS does not yield as much gain over the panel design.

The improvements of iterated ACS is useful given relatively small $m$, if positively size-biased sampling is difficult to achieve, e.g. due to a lack of understanding of the relevant risk factors, or a lack of frame data that can be used to effectively vary the initial sample inclusion probability even though the relevant factors are known. Together, the panel and iterated ACS designs seem to complement each other in different settings, offering helpful choices across a wider range of situations than each on its own.

## BIBLIOGRAPHIC NOTES

Thompson (1990) proposes the modified HT-estimator for ACS, which is presented as the strategy $(G, \hat{\theta}^*_{HT})$ here. Thompson (1990) proposes similarly also a modified HH-type estimator based on the multiplicity weights, where an edge grid is used for estimation only if it is selected in $s_0$ directly. This modified HH-type estimator is the same as the unmodified HH-type estimator under BIGS from $\mathcal{B}^*$ in Section 4.1.1.

Zhang and Oguz-Alper (2020) discuss the HTE for various BIGS associated with ACS from $G$, based on the recognition that one can either modify the sampling method or the estimator in a sampling strategy. Admitting all the observational links under ACS, one can modify the HT or HH-type estimators, as originally proposed by Thompson (1990). Using $\mathcal{B}^*$ or $\mathcal{B}^\dagger$ that modifies the observational links under ACS to ensure the ancestry knowledge, one can use the unmodified HT or HH-type estimator.

The two-stage ACS example is taken from Zhang and Oguz-Alper (2020). The estimators other than $\hat{\theta}_{z\alpha 1}$ are also considered by Thompson (1991).

The designs for cross-sectional and change estimation of case prevalence in Sections 4.2 and 4.3 are developed by Zhang (2020).

# CHAPTER 5

# Snowball sampling

Repeated incident OP enlarges a sample of nodes by their out-of-sample successors in a manner which resembles the rolling of a snowball. $T$-wave *snowball sampling (TSBS)* is a breadth-first graph sampling method. We consider the estimation of graph totals of finite-order induced motifs by $T$SBS.

## 5.1 $T$-WAVE SNOWBALL SAMPLING

$T$SBS from $G = (U, A)$ is given by applying the $T$-wave incident OP to an initial sample $s_0$, as defined in Section 1.3.3. We assume $F = U$ here. The sample graph $G_s = (U_s, A_s)$ follows from (1.3). For digraphs, the reference set and the sample edges are given by

$$s_{\text{ref}} = s \times U \qquad \text{and} \qquad A_s = \bigcup_{i \in s} \bigcup_{j \in \alpha_i} A_{ij}$$

where $s$ is the seed sample. For undirected graphs, we have

$$s_{\text{ref}} = s \times U \cup U \times s \qquad \text{and} \qquad A_s = \bigcup_{i \in s} \bigcup_{j \in \alpha_i} (A_{ij} \cup A_{ji})$$

### 5.1.1 Inclusion probabilities of nodes and edges

By way of introduction consider the inclusion probabilities $\pi_{(i)}$ and $\pi_{(i)(j)}$ of nodes in $U_s$, and $\pi_{(ij)}$ and $\pi_{(ij)(hl)}$ of edges in $A_s$.

DOI: 10.1201/9781003203490-5

The wave samples $s_0, ..., s_T$ are disjoint under $T$SBS. A given sample node $i$ can only belong to one particular $s_t$, where $t \leq T$. For any $i \in U$, let $\beta_i^0 = \{i\}$, and let its $t$-th *wave ancestors* be

$$\beta_i^t = \beta(\beta_i^{t-1}) \setminus \bigcup_{r=0}^{t-1} \beta_i^r \quad \text{for} \quad t > 0$$

which consists of the nodes that would lead to $i$ in $t$ waves but not sooner, and $\beta_i^0, \beta_i^1, ..., \beta_i^T$ are disjoint. The $T$SBS ancestors of node $i$ is given by

$$F_i = \bigcup_{t=0}^{T} \beta_i^t$$

For any $i, j \in U$, we have

$$\pi_{(i)} = 1 - \bar{\pi}_{F_i} \tag{5.1}$$

$$\pi_{(i)(j)} = 1 - \bar{\pi}_{F_i} - \bar{\pi}_{F_j} + \bar{\pi}_{F_i \cup F_j}$$

Moreover, for any $i \neq j \in U$ and $h \neq l \in U$, we have

$$\pi_{(ij)} = 1 - \bar{\pi}_{F_{ij}} \tag{5.2}$$

$$\pi_{(ij)(hl)} = 1 - \bar{\pi}_{F_{ij}} - \bar{\pi}_{F_{hl}} + \bar{\pi}_{F_{ij} \cup F_{hl}}$$

where the ancestor set $F_{ij}$ is given by

$$F_{ij} = \bigcup_{t=0}^{T-1} \beta_i^t$$

in digraphs, and it is given by

$$F_{ij} = \bigcup_{t=0}^{T-1} (\beta_i^t \cup \beta_j^t)$$

in undirected graphs where $(ij) = (ji)$ by definition.

Figure 5.1 shows the sample graph $G_s = (U_s, A_s)$ by 2SBS given $s_0 = \{3, 4\}$, from the population graph $G = (U, A)$. The 1st and 2nd wave samples are $s_1 = \{8, 9, 10\}$ and $s_2 = \{1, 5, 7\}$. The seed sample is $s = \{3, 4, 8, 9, 10\}$. The node inclusion probabilities $\pi_{(i)}$ are given by (5.1), where there are 5 distinct values over $U$, and the edge inclusion probabilities $\pi_{(ij)}$ by (5.2), where there are 4 distinct values over $A$. These can be verified by enumeration over all the 45 possible initial samples given $|s_0| = 2$. For example, the node 3 has $\beta_3^0 = \{3\}$, $\beta_3^1 = \{4\}$ and $\beta_3^2 = \{8, 9, 10\}$, such that we have $F_3 = \{3, 4, 8, 9, 10\}$. Under SRS of $s_0$, the exclusion probability of $F_3$ from $s_0$ is $\binom{5}{2}/\binom{10}{2} = 2/9$, such that $\pi_{(3)} = 1 - 2/9 = 7/9$, whereas $\pi_3 = \Pr(3 \in s_0) = 1/5$.

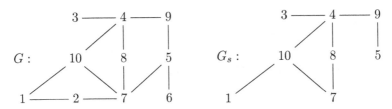

**Figure 5.1** 2SBS from $G$ given $s_0 = \{3, 4\}$

## 5.1.2 Arbitrary $M$ with $|M| \geq 2$ given $s_{\text{ref}} = s \times U \cup U \times s$

Let $M = \{i_1, i_2, ..., i_q\}$ with $|M| = q$. To identify the induced motif $\kappa = [M]$, there can be at most one node in $M$ that belongs to the last wave sample $s_T$. Let $M^{(i_h)} = M \setminus \{i_h\}$ be the subset obtained by dropping $i_h$ from $M$, for $h = 1, ..., q$. We have

$$\pi_{(\kappa)} = \Pr\left(M^{(i_1)} \subseteq s \text{ or } \cdots \text{ or } M^{(i_q)} \subseteq s \text{ or } M \subseteq s\right)$$

$$= \sum_{h=1}^{q} \Pr\left(M^{(i_h)} \subseteq s\right) - (q-1)\Pr(M \subseteq s) \qquad (5.3)$$

This follows from applying the inclusion-exclusion calculus of the probability of union of events. We have $\left(M^{(i_h)} \subseteq s\right) \cap \left(M \subseteq s\right) = \left(M \subseteq s\right)$ for any $i_h$, and $\left(M^{(i_h)} \subseteq s\right) \cap \left(M^{(i_l)} \subseteq s\right) = \left(M \subseteq s\right)$ for $i_h \neq i_l$, such that all the terms are proportional to $\Pr\left(M \subseteq s\right)$ except $\sum_{h=1}^{q} \Pr\left(M^{(i_h)} \subseteq s\right) + \Pr\left(M \subseteq s\right)$. The sum of the former terms, $\left(1 - (q+1)\right)\Pr\left(M \subseteq s\right)$, can be obtained from matching them to the corresponding terms in the identity $(1-1)^{q+1} = 0$.

Next, we have

$$\Pr\left(M \subseteq s\right) = \sum_{L \subseteq M} (-1)^{|L|} \bar{\pi}(L)$$

$$\bar{\pi}(L) = \Pr(L \cap s = \emptyset) = \bar{\pi}_{F_L} = \sum_{D \subseteq F_L} (-1)^{|D|} \pi_D$$

where $\bar{\pi}(L)$ is the seed-sample exclusion probability including $\bar{\pi}(\emptyset) = 1$, and $F_L = \bigcup_{i \in L} F_i$ with $F_i$ in (5.1), and $\pi_D$ is the joint inclusion probability in the initial sample $s_0$. Similarly for $\Pr\left(M^{(i_h)} \subseteq s\right)$, where $h = 1, ..., q$.

For $M \subset U$ and $M' \subset U$, joint observation of $\kappa = [M]$ and $\kappa' = [M']$ requires at most one node $i$ in $s_T$ if $i \in M \cap M'$, or at most two nodes $i, j$ in $s_T$ if $i \in M \setminus M'$ and $j \in M' \setminus M$.

Let $\tilde{M} = M \cup M'$. Define subset $\tilde{M}^{(i)}$ for each $i \in M \cap M'$, and $\tilde{M}^{(ij)}$ for each pair of $i \in M \setminus M'$ and $j \in M' \setminus M$. The joint inclusion probability $\pi_{(\kappa\kappa')}$ is the probability that at least one of these subsets of $\tilde{M}$ is in the seed sample $s$, which can be obtained similarly as (5.3).

### 5.1.3 Arbitrary $M$ with $|M| \geq 2$ using $s_{\text{ref}}^* = s \times s$

Incident OP under $T$SBS implies induced OP in $s$. By dropping the last wave sample $s_T$, one can ensure that an induced motif $[M]$ is observed if $M \subset s$. Let $G_s^* = (U_s^*, A_s^*)$ be the sample graph obtained from dropping $s_T$, where

$$U_s^* = U_s \setminus s_T = s \quad \text{and} \quad A_s^* = A_s \cap s_{\text{ref}}^* \quad \text{and} \quad s_{\text{ref}}^* = s \times s$$

containing all the edges between any $i, j \in s$ in $G$. That is, $G_s^*$ is the sample graph obtained from $s$ by induced OP directly. It follows that $\pi_{(\kappa)}^*$ with respect to $s_{\text{ref}}^*$ is given by

$$\pi_{(\kappa)}^* = \Pr\left(M \subseteq s\right) \tag{5.4}$$

For $M \neq M' \subset U$, joint observation of $\kappa = [M]$ and $\kappa' = [M']$ requires simply $M \cup M' \subseteq s$, such that $\pi_{(\kappa\kappa')}^* = \Pr(M \cup M' \subseteq s)$ with respect to $s_{\text{ref}}^*$.

### 5.1.4 Illustration of two strategies

For sampling from undirected graphs, using the HTE based on $\pi_{(\kappa)}$ by (5.3) with respect to $s_{\text{ref}} = s \times U \cup U \times s$ is a different strategy than using the modified HTE based on $\pi_{(\kappa)}^*$ by (5.4) with respect to $s_{\text{ref}}^* = s \times s$. One may expect a loss of efficiency by using the modified HTE, while the HTE requires more computation.

Table 5.1 Inclusion probability of selected triad motif $\kappa = [M]$ in Figure 5.2

| $M$ | $\{1,2,3\}$ | $\{1,2,4\}$ | $\{1,3,4\}$ | $\{2,3,4\}$ | $\{1,2,5\}$ |
|---|---|---|---|---|---|
| $\pi_{(\kappa)}$ | 0.923 | 0.853 | 0.832 | 0.853 | 0.867 |
| $\pi_{(\kappa)}^*$ | 0.566 | 0.266 | 0.203 | 0.255 | 0.622 |

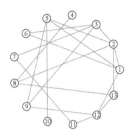

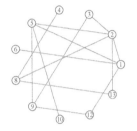 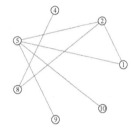

**Figure 5.2**  2SBS from $G$ (top) given $s_0 = \{4, 5, 10\}$, sample graphs $G_s$ (bottom left) and $G_s^*$ (bottom right)

Figure 5.2 illustrates 2SBS from $G = (U, A)$, where the two sample graphs $G_s$ and $G_s^*$ follow from $s_0 = \{4, 5, 10\}$ by SRS. We have $s_1 = \{1, 2, 8, 9\}$, $s_2 = \{3, 6, 12, 13\}$ and $s = \{1, 2, 4, 5, 8, 9, 10\}$. Table 5.1 lists the inclusion probabilities of 5 selected triad motifs in $G$ and $G_s^*$, respectively given by (5.3) and (5.4). Note that every motif has a higher inclusion probability in $G_s$ than $G_s^*$.

Table 5.2 shows the HTE of four 3rd-order graph totals $\hat{\theta}_{3,h}$, which are the numbers of triads of size $h = 0, 1, 2, 3$, as defined in Section 1.2.4. Their expectations and standard errors are also given, where the expectations are the true totals in $G$, because the HTE is unbiased. Similarly for the modified HTEs based on $\pi_{(\kappa)}^*$, which are also unbiased but less efficient than the HTE.

Table 5.2  Estimation of third-order graph totals in Figure 5.2

| HTE | $\hat{\theta}_{3,0}$ | $\hat{\theta}_{3,1}$ | $\hat{\theta}_{3,2}$ | $\hat{\theta}_{3,3}$ |
|---|---|---|---|---|
| Estimate | 96.3 | 89.3 | 26.3 | 2.5 |
| Expectation | 121 | 123 | 40 | 2 |
| Std. error | 23.0 | 18.6 | 7.0 | 0.8 |
| Mod. HTE | $\hat{\theta}_{3,0}^*$ | $\hat{\theta}_{3,1}^*$ | $\hat{\theta}_{3,2}^*$ | $\hat{\theta}_{3,3}^*$ |
| Estimate | 59.1 | 63.2 | 19.2 | 1.6 |
| Expectation | 121 | 123 | 40 | 2 |
| Std. error | 78.7 | 49.9 | 15.0 | 1.2 |

## 5.2  STRATEGIES FOR $T$SBS

To implement the HTE by (5.3), one needs to know all the relevant ancestor sets $F_i$, which is generally not possible given only the sample graph. It is also more convenient to work with the ancestors of a given motif directly, rather than working through the ancestors of each node in the motif.

### 5.2.1  Distances to a motif

Let $\varphi_{ij}$ be the geodesic distance from node $i$ to node $j$ in $G$. Take Figure 5.3, where there are no edges between any of these 8 nodes and the rest of the graph. We have among others $\varphi_{23} = \varphi_{67} = 1$, $\varphi_{14} = \varphi_{78} = 2$ and $\varphi_{16} = \varphi_{35} = \infty$.

Figure 5.3  Illustration of distances to a motif

The *geodesic distance from node $i$ to motif* $\kappa$ is the number of waves it takes from $i$ to reach *any* node in $M(\kappa)$, denoted by $\varphi_{i,\kappa}$. We have $\varphi_{i,\kappa} = 0$ if $i \in M(\kappa)$, and

$$\varphi_{i,\kappa} = \min_{j \in M(\kappa)} \varphi_{ij}$$

if $i \notin M(\kappa)$. For instance, take the 2-star motif of $M = \{6, 7, 8\}$ in Figure 5.3, we have $\varphi_{6,\kappa} = \varphi_{7,\kappa} = \varphi_{8,\kappa} = 0$, $\varphi_{5,\kappa} = 1$ and $\varphi_{i,\kappa} = \infty$ for any other node $i$.

The *radius distance from node $i$ to motif* $\kappa$ is the number of waves it takes from $i$ to reach *all* the nodes in $M(\kappa)$, denoted by $\lambda_{i,\kappa}$. For any $i \in U$, we have

$$\lambda_{i,\kappa} = \max_{j \in M(\kappa)} \varphi_{ij}$$

For $M = \{6, 7, 8\}$ in Figure 5.3, we have $\lambda_{6,\kappa} = 1$, $\lambda_{7,\kappa} = \lambda_{8,\kappa} = 2$, $\lambda_{5,\kappa} = 2$ and $\lambda_{i,\kappa} = \infty$ for any other node $i$.

The *observation distance from node $i$ to motif* $\kappa$ is the number of waves it takes from $i$ to observe the motif $\kappa$, denoted by $d_{i,\kappa}$. For $\kappa$ of $M = \{6, 7, 8\}$ in Figure 5.3, we have $d_{6,\kappa} = d_{7,\kappa} = d_{8,\kappa} = 2$ as well as $d_{5,\kappa} = 2$. Whereas for $\kappa$ of $M = \{1, 2, 3, 4\}$, we have $d_{1,\kappa} = d_{3,\kappa} = d_{4,\kappa} = 3$ and $d_{2,\kappa} = 2$. The calculus of observation distance to any induced motif is summarised below.

**Lemma 5.1.** *For any $\kappa \in \Omega$ and $i \in U$, if the nodes $M(\kappa)$ are connected in $G$, then*

$$d_{i,\kappa} = \begin{cases} \lambda_{i,\kappa} & \textit{if } |\arg\max_{j \in M} \varphi_{ij}| = 1 \\ 1 + \lambda_{i,\kappa} & \textit{otherwise} \end{cases}$$

*Proof.* If $|\arg\max_{j \in M} \varphi_{ij}| = 1$, then there is only one node in $M$, denoted by $j_0$, which requires $\lambda_{i,\kappa}$ waves from $i$. All the other nodes in $M$ must be observed before $j_0$, which allows one to observe any edge between them by the last wave at the latest when $j_0$ is observed. If there are more than one node that requires $\lambda_{i,\kappa}$ waves from $i$, then an additional wave is needed to observe the edges among them. □

**Lemma 5.2.** *For any $\kappa \in \Omega$ and $i \in U$, if there exists a single node in $M(\kappa)$, which is unconnected to $i$ in $G$, then*

$$d_{i,\kappa} = 1 + \max_{j \in M(\kappa;i)} \varphi_{ij}$$

*where $M(\kappa; i)$ are the nodes in $M(\kappa)$, which are connected to $i$.*

*Proof.* All the nodes $M(\kappa; i)$ are reached from $i$ by $\max_{j \in M(\kappa;i)} \varphi_{ij}$ waves and not earlier. An additional wave is needed to confirm that the last node is unconnected to $M(\kappa; i)$, as well as to ensure that one observes all the edges among $M(\kappa; i)$. □

Finally, let the *diameter* of a motif $\kappa$ be

$$\varphi_\kappa = \max_{i,j \in M(\kappa)} \varphi_{ij}$$

The subgraph $G(M)$ must be connected if the motif $[M]$ has a finite diameter. Let the *observation diameter* of the motif $\kappa$ be

$$\zeta_\kappa = \max_{i \in M(\kappa)} d_{i,\kappa}$$

By Lemma 5.1, given any $M(\kappa)$ with $\varphi_\kappa < \infty$, we have

$$\zeta_\kappa \leq 1 + \varphi_\kappa$$

For instance, in Figure 5.3, $\varphi_\kappa = \zeta_\kappa = 2$ for $M(\kappa) = \{6,7,8\}$, $(\varphi_\kappa, \zeta_\kappa) = (1,2)$ for $M(\kappa) = \{5,6,7\}$, or $(\varphi_\kappa, \zeta_\kappa) = (2,3)$ for $M(\kappa) = \{1,2,3,4\}$.

## 5.2.2 Using $F_\kappa$

The node $i$ in $G$ is a *TSBS ancestor* of motif $\kappa$ if $d_{i,\kappa} \leq T$. Suppose the ancestor set $F_\kappa$ satisfies (3.1) for all $\kappa \in \Omega$. For any sample $\kappa$ in undirected graphs, additional waves of OP may be needed to identify $F_\kappa$ completely. One cannot directly observe the out-of-sample ancestors by additional waves of incident OP in digraphs.

**Lemma 5.3.** *Given any sample motif $\kappa$ with $\varphi_\kappa < \infty$ in undirected graphs, if $|M| > 1$ then one needs at most $T - 1$ waves of incident OP from $M$ to observe all the TSBS ancestors $F_\kappa$, if $|M| = 1$ then one needs at most $T$ waves of incident OP from $M$.*

*Proof.* Applying $T - 1$ waves of incident OP to $M$, as if $s_0 = M$, would identify *all* the nodes $\{i \in U : \varphi_{i,\kappa} \leq T-1\}$. By Lemma 5.1, if $|M| > 1$, then $F_\kappa$ must be a subset of them. Whereas if $|M| = 1$, then applying at most $T$ waves of incident OP to $M$ would identify all the nodes $F_\kappa$. □

Having identified all the TSBS ancestors of each $\kappa$ in $\Omega_s$ by Lemma 5.3, one can fix an associated BIGS by (3.2) and apply the IWE (2.2) accordingly.

What Lemma 5.3 provides is an upper bound. Take the triangle $\kappa$ with nodes $M = \{i_1, i_2, i_3\}$ in Figure 5.4 for illustration. Suppose 3SBS, where $F_\kappa$ is the union of $\{i_1, i_2, i_3, j_2, j_3\}$ with $d_{i,\kappa} = 2$ and

Figure 5.4  Illustration of strategies for $TSBS$

$\{j_1, h_1, h_2\}$ with $d_{i,\kappa} = 3$. Given $\kappa \in \Omega_s$, we need at most 2 more waves to identify all its 3SBS ancestors by Lemma 5.3. Indeed, in this case we only need at most 1 more wave from the moment $\kappa$ is observed, which can be verified by enumerating all the possibilities where $|s_0 \cap F_\kappa| = 1$. For instance, we need one more wave to observe $h_2$ starting from only $h_1 \in s_0$.

### 5.2.3  Using known subset of $F_\kappa$

One can limit $\beta_\kappa$ in (3.2) to a known subset of $F_\kappa$ and apply the corresponding strategy BIGS-IWE without additional waves of OP, such as by the two results below.

**Lemma 5.4.** *Provided $\zeta_\kappa < \infty$ for any $\kappa \in \Omega$, fixing $\beta_\kappa^* = M(\kappa)$ in (3.2) yields an unbiased strategy BIGS-IWE for $\theta$ by (1.2) under $TSBS$ with $T = \max_{\kappa \in \Omega} \zeta_\kappa$.*

A sample motif $\kappa \in \Omega_s$ would be ineligible for estimation given $\beta_\kappa^* = M(\kappa)$, if $s_0 \cap \beta_\kappa^* = \emptyset$. When the sampling variance of $TSBS$ determined by Lemma 5.4 is too large to be acceptable, one may like to increase $T$. This raises the need to update the associated BIGS for $TSBS$, where $T > \max_{\kappa \in \Omega} \zeta_\kappa$. Let

$$\beta_\kappa^t(M) = \{i \notin M(\kappa) : \varphi_{i,\kappa} \le t\}$$

contain all the nodes *outside of* $M(\kappa)$, which have maximum geodesic distance $t$ to $\kappa$. That is, starting from any $i \in \beta_\kappa^t(M)$, it takes at most $t$ waves of incident OP in $G$ to reach $M(\kappa)$. Given $T > \max_{\kappa \in \Omega} \zeta_\kappa$, the nodes in $\beta_\kappa^t(M)$ may be added to $\beta_\kappa$ in (3.2), for $t \ge 1$, according to the result below.

**Lemma 5.5.** *Provided $\zeta_\kappa < \infty$ for any $\kappa \in \Omega$, fixing $\beta_\kappa^* = M(\kappa) \cup \beta_\kappa^t(M)$ in (3.2), for $t \ge 1$, yields an unbiased strategy BIGS-IWE for $\theta$ by (1.2) under $TSBS$ with*

$$T = \max_{\kappa \in \Omega} \varphi_\kappa + 2t$$

*Proof.* Any motif $\kappa$ is observed after at most $\zeta_\kappa + t$ waves starting from any node in $\beta_\kappa^* = M(\kappa) \cup \beta_\kappa^t(M)$. In the case $\zeta_\kappa = 1 + \varphi_\kappa$, all the nodes $M(\kappa)$ must have been observed at wave $\zeta_\kappa + t - 1$, so that $\beta_\kappa^1(M)$ are already observed after $\zeta_\kappa + t = 1 + \varphi_\kappa + t$ waves. It remains only to observe all the geodesics to $\beta_\kappa^t(M) \backslash \beta_\kappa^1(M)$ starting from $\beta_\kappa^1(M)$, which requires at most $t - 1$ waves. Otherwise, in the case $\zeta_\kappa = \varphi_\kappa$, there is at least one node $j \in M(\kappa)$, which is first observed at wave $\zeta_\kappa$ starting from any node in $M(\kappa)$. Up to $t$ additional waves may be needed to observe all the nodes outside $M(\kappa)$, which can lead to $j$ in $t$ waves. Thus, in either case, $\beta_\kappa^*$ can always be identified as $T$SBS ancestors based on the sample graph $G_s$ in which $\kappa$ is observed. □

Take $M(\kappa) = \{i_1, i_2, i_3\}$ in Figure 5.4, with $(\varphi_\kappa, \zeta_\kappa) = (1, 2)$. First, $\beta_\kappa^* = M(\kappa)$ for 2SBS by Lemma 5.4, which excludes the other two 2SBS ancestors $j_2, j_3$.

Next, $\beta_\kappa^* = M(\kappa) \cup \beta_\kappa^1(M)$ for 3SBS by Lemma 5.5, where $\beta_\kappa^1(M) = \{j_1, j_2, j_3\}$, which excludes $h_1, h_2 \in F_\kappa$.

Finally, $\beta_\kappa^* = M(\kappa) \cup \beta_\kappa^1(M)$ remains the same for 4SBS by Lemma 5.5. For instance, the nodes ★, $h_1$ and $h_2$ all belong to $\beta_\kappa^2(M) \backslash \beta_\kappa^1(M)$. However, when only ★ $\in s_0$, we would observe $\kappa$ under 4SBS but not $(j_2 h_2)$, such that $h_2$ could not be identified as a 4SBS ancestor.

## 5.2.4 Using subset of $F_{\kappa|s}$

One can limit $\beta_\kappa$ in (3.2) to a subset of ancestors $F_{\kappa|s}$, which are identified from the sample graph $G_s$, and apply the corresponding strategy BIGS-IWE without additional waves of OP. Inference needs to be made conditional on the chosen strategy.

Given the sample graph $G_s$ observed under $T$SBS from $G$, let $d_{i,\kappa}(G_s)$ be the *sample observation distance* from $i$ to given motif $\kappa$ in $G_s$, where $d_{i,\kappa}(G_s) \geq d_{i,\kappa}(G)$ generally. Given any $\kappa \in \Omega_s$, the observed sample $T$SBS ancestors are

$$F_{\kappa|s} = \{i \in U_s : d_{i,\kappa}(G_s) \leq T\} \tag{5.5}$$

Note that for each $i \in F_{\kappa|s}$, there exists at least a path corresponding to $d_{i,\kappa}(G_s)$ in the sample subgraph induced by $F_{\kappa|s}$, denoted by $G_s(F_{\kappa|s})$. This is because any node on the path giving rise to

$d_{i,\kappa}(G_s)$ must have a shorter sample observation distance to $\kappa$ than $i$, which means that it must belong to $F_{\kappa|s}$.

There exists an *ancestor network* of $\kappa$, denoted by $F^*_{\kappa|s} \subseteq F_{\kappa|s}$, in the sense that $F^*_{\kappa|s}$ (but not necessarily $F_{\kappa|s}$) can all be identified as the $T$SBS ancestors whenever $s_0 \cap F^*_{\kappa|s} \neq \emptyset$. This requires one to be able to observe the induced subgraph $G_s(F^*_{\kappa|s})$ under $T$SBS starting from any $i \in F^*_{\kappa|s}$. In other words, $F^*_{\kappa|s}$ can be given as

$$F^*_{\kappa|s} = \{i \in F_{\kappa|s} : d_{i,\tau}(G_s) \leq T, \ \tau = [G_s(F^*_{\kappa|s})]\} \qquad (5.6)$$

Generally, the ancestor network $F^*_{\kappa|s}$ may differ cross all the different sample graphs where $\kappa$ is observed; nor do we always have $\beta^*_\kappa \subseteq F^*_{\kappa|s}$ or $\beta^*_\kappa \subseteq F_{\kappa|s}$, where $\beta^*_\kappa$ is fixed by Lemma 5.5.

Take $M(\kappa) = \{i_1, i_2, i_3\}$ in Figure 5.4, for which $\beta^*_\kappa$ by Lemma 5.4 or 5.5 has been given before for $T$SBS with $T = 2, 3, 4$.

First, for 2SBS, we have $F_{\kappa|s} = F^*_{\kappa|s} = M(\kappa) \cup \{j_2, j_3\}$ by (5.5) and (5.6), which is also $F_\kappa$ in this case.

Next, for 3SBS, if $s_0 \cap \beta_\kappa \subseteq \{j_1, h_1\}$, then

$$F_{\kappa|s}(j_1, h_1) = F^*_{\kappa|s}(j_1, h_1) = M(\kappa) \cup \{j_2, j_3\} \cup \{j_1, h_1\}$$

where $(j_2 h_2)$ is unobserved. If $s_0 \cap \beta_\kappa = \{h_2\}$, then

$$F_{\kappa|s}(h_2) = F^*_{\kappa|s}(h_2) = M(\kappa) \cup \{j_2, j_3\} \cup \{h_2\}$$

where neither $(i_1 j_1)$ nor $(j_3 h_1)$ is observed. We would obtain

$$F_{\kappa|s} = M(\kappa) \cup \{j_2, j_3\} \cup \{j_1, h_1, h_2\} = F_\kappa$$

in all the other situations where $\kappa$ is observed under 3SBS. For an ancestor network, one can e.g. let $F^*_{\kappa|s} = F^*_{\kappa|s}(h_2)$ by removing $\{j_1, h_1\}$ or one can let $F^*_{\kappa|s} = F^*_{\kappa|s}(j_1, h_1)$ by removing $h_2$.

Finally, for 4SBS, $F_\kappa$ now includes also the nodes like ★, in addition to the 3SBS ancestors. We note the following.

- If $F^*_{\kappa|s}$ by (5.6) includes the ★-node in Figure 5.4, then it must exclude $h_2$ and the other nodes adjacent to $j_2$ or $h_1$. If $F^*_{\kappa|s}$ by (5.6) includes $h_2$ and other nodes adjacent to $j_2$, then it must exclude the ★-node and the other nodes adjacent to $h_1$.

- If $M(\kappa) \cup \{j_1, j_2, j_3, h_1\}$ intersects $s_0$, then $F_{\kappa|s}$ would include $h_2$ and ★, as well as the nodes adjacent to $j_2$ and $h_1$, although not all of them can be included in $F^*_{\kappa|s}$.

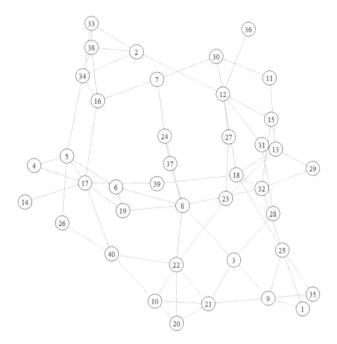

Figure 5.5  A population graph with $|U| = 40$ and $|A| = 72$

## 5.3  ILLUSTRATIONS

Figure 5.5 shows a population graph $G$ of 40 nodes and 72 edges. Consider the motifs illustrated in Figure 1.2: node ($\mathcal{K}_1$), 2-clique ($\mathcal{K}_2$), 2-star ($\mathcal{S}_2$), triangle or 3-clique ($\mathcal{K}_3$), 4-clique ($\mathcal{K}_4$), 4-cycle ($\mathcal{C}_4$), 3-star ($\mathcal{S}_3$) and 3-path ($\mathcal{P}_3$). Their totals in Figure 5.5 are

$$(\theta_{\mathcal{K}_1}, \theta_{\mathcal{K}_2}, \theta_{\mathcal{S}_2}, \theta_{\mathcal{K}_3}, \theta_{\mathcal{K}_4}, \theta_{\mathcal{C}_4}, \theta_{\mathcal{S}_3}, \theta_{\mathcal{P}_3})$$
$$= (40, 179, 72, 19, 3, 7, 141, 408)$$

The diameters $\varphi_\kappa$ and the observation diameters $\zeta_\kappa$ of these motifs are given at the top of Table 5.3. First, $\zeta_\kappa$ waves of SBS is required by Lemma 5.4 for an IWE, such as the HTE $\hat{\theta}_y$ or the multiplicity estimator $\hat{\theta}_{z\beta}$ for associated BIGS using $\beta_\kappa^* = M(\kappa)$, whereas one may need up to $\zeta_\kappa$ additional waves for the HH-type estimator $\hat{\theta}_{z\alpha\gamma}$ using the PIDA weights (2.10). Next, $\varphi_\kappa + 2t$ waves of SBS is required by Lemma 5.5 for $\hat{\theta}_y$ and $\hat{\theta}_{z\beta}$ under BIGS using

$\beta_\kappa^* = M(\kappa) \cup \beta_\kappa^t(M)$ with $t \geq 1$, whereas up to $\zeta_\kappa + t$ additional waves of observation may be needed for $\hat{\theta}_{z\alpha\gamma}$.

For an example of sample-dependent strategy using $F_{\kappa|s}$, take $M(\kappa) = \{2, 34, 38\}$ in the top-left corner of Figure 5.5, which is an occurrence of $\mathcal{K}_3$. Under 2SBS, we have $F_{\kappa|s} = F_\kappa = M(\kappa) \cup \{33\}$ whenever $\kappa \in \Omega_s$, whereas $\beta_\kappa^* = M(\kappa)$ by Lemma 5.4. Under 3SBS, we have $F_{\kappa|s} = F_\kappa = M(\kappa) \cup \{33\} \cup \{5, 12, 16\}$ whenever $\kappa \in \Omega_s$, which is the same as $\beta_\kappa^*$ by Lemma 5.5 in this case.

### 5.3.1 An example

In Figure 5.6, the initial sample $s_0 = \{3, 12\}$ by SRS is marked as $T$SBS with $T = 0$; the other observed sample graphs are marked as $T = 1, 2, 3, 4$. The sample graph by 4SBS includes all the nodes in $G$ but not all the edges.

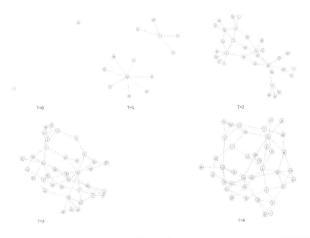

**Figure 5.6** Initial $s_0 = \{3, 12\}$, sample graphs by $T$SBS, $T \leq 4$

To illustrate some of the computational details, let 4-cycle $\mathcal{C}_4$ be the motif of interest. The graph total of $\mathcal{C}_4$ is 7.

By Lemma 5.4, we have $\beta_\kappa^* = M(\kappa)$ for 2SBS, where $\zeta_\kappa = 2$ for $\mathcal{C}_4$. The details required for $\hat{\theta}_y$, $\hat{\theta}_{z\beta}$ and $\hat{\theta}_{z\alpha\gamma}$ are given in the upper part of Table 5.4. The motif $A$ with nodes $M(A) = \{3, 8, 21, 22\}$ is observed from node 3, and the motifs $B$ and $C$ from node 12, where $M(B) = \{12, 13, 18, 31\}$ and $M(C) = \{12, 15, 18, 32\}$. Two more waves are needed to apply $\hat{\theta}_{z\alpha\gamma}$, where the relevant $|\alpha_i^*|$ in

Table 5.3  Diameter and observation diameter of motif, number of waves $T$ required for corresponding IWE under associated BIGS with specified $\beta_\kappa^*$ in (3.2)

| $\beta_\kappa^*$ | | $\mathcal{K}_1$ | $\mathcal{K}_2$ | $\mathcal{S}_2$ | $\mathcal{K}_3$ | $\mathcal{K}_4$ | $\mathcal{C}_4$ | $\mathcal{S}_3$ | $\mathcal{P}_3$ |
|---|---|---|---|---|---|---|---|---|---|
| | $\varphi_\kappa$ | 0 | 1 | 2 | 1 | 1 | 2 | 2 | 3 |
| | $\zeta_\kappa$ | 0 | 1 | 2 | 2 | 2 | 2 | 3 | 3 |
| $M(\kappa)$ | $T$ for $\hat\theta_{z\beta}$ | 0 | 1 | 2 | 2 | 2 | 2 | 3 | 3 |
| | $T$ for $\hat\theta_{z\alpha\gamma}$ | 0 | 2 | 4 | 3 | 3 | 4 | 5 | 6 |
| $M(\kappa)\cup\beta_\kappa^1(M)$ | $T$ for $\hat\theta_{z\beta}$ | 2 | 3 | 4 | 3 | 3 | 4 | 4 | 5 |
| | $T$ for $\hat\theta_{z\alpha\gamma}$ | 3 | 5 | 7 | 6 | 6 | 7 | 8 | 9 |
| $M(\kappa)\cup\beta_\kappa^2(M)$ | $T$ for $\hat\theta_{z\beta}$ | 4 | 5 | 6 | 5 | 5 | 6 | 6 | 7 |
| | $T$ for $\hat\theta_{z\alpha\gamma}$ | 6 | 8 | 10 | 9 | 9 | 10 | 11 | 12 |

$\mathcal{B}^*$ are given in the last column of Table 5.4, such as

$$\{|\alpha_i^*| : i \in \beta_B^*\} = \{|\alpha_{12}^*|, |\alpha_{13}^*|, |\alpha_{18}^*|, |\alpha_{31}^*|\} = \{2, 2, 3, 1\}$$

for the motif $B$.

Under SRS with $m = |s_0|$, the inclusion probability of $\kappa$ is

$$\pi_{(\kappa)} = 1 - \binom{N - |\beta_\kappa^*|}{m} / \binom{N}{m} \equiv 0.1923$$

where $N = 40$, $m = 2$ and $|\beta_\kappa^*| \equiv 4$ for any motif $\kappa$ that is $\mathcal{C}_4$. By (2.6), we have $\hat{\theta}_y = 3/0.1923 = 15.6$. For $\hat{\theta}_{z\beta}$ by (2.8), we have $z_3 = 1/4$ from $\alpha_3^* = \{A\}$ and $z_{12} = 1/4 + 1/4$ from $\alpha_{12}^* = \{B, C\}$, such that $\hat{\theta}_{z\beta} = (3/4)/(2/40) = 15$, where $\pi_i \equiv 2/40$. For $\hat{\theta}_{z\alpha 1}$ by (2.10), we have $z_3 = 1/4$ from $\alpha_3^*$ and $z_{12} = 0.5/2.33 + 0.5/2.33 = 0.43$ from $\alpha_{12}^*$, such that $\hat{\theta}_{z\alpha 1} = 13.6$.

By Lemma 5.5, setting $\beta_\kappa^* = M(\kappa) \cup \beta_\kappa^1(M)$ with $t = 1$ is feasible under 4SBS for $\mathcal{C}_4$ with $\varphi_\kappa = 2$. The details are given in the lower part of Table 5.4. The motif $A$ is observed from node 3, and the motifs $\{B, C, D\}$ from node 12 where, compared to 2SBS, the extra motif $D$ with $M(D) = \{13, 18, 29, 32\}$ is observed via node 18 obtained at the 1st wave (Figure 5.6). The cardinality of the ancestor set $\beta_\kappa^*$ is given in Table 5.4, which is 15, 16, 14 or 12 for $\kappa = A, B, C$ or $D$. More waves of observation are needed to apply $\hat{\theta}_{z\alpha\gamma}$, so that it is not applicable under 4SBS here, and the relevant $|\alpha_i^*|$ are omitted in Table 5.4.

The probability $\pi_{(\kappa)}$ is 0.6154, 0.6462, 0.5833 and 0.5154 for $\kappa = A, B, C$ and $D$, which are much higher than $\pi_{(k)} \equiv 0.1923$ under 2SBS above. By (2.6), we have $\hat{\theta}_y = 6.83$. For $\hat{\theta}_{z\beta}$ by (2.8), we have $z_3 = 1/15$ from $\alpha_3^* = \{A\}$ and $z_{12} = 1/16 + 1/14 + 1/12$ from $\alpha_{12}^* = \{B, C, D\}$, such that $\hat{\theta}_{z\beta} = 5.68$. Both these two estimates are much closer to the graph total $\theta = 7$ than those under 2SBS, with only one extra motif $D$.

## 5.3.2 Some results

Consider SBS up to 4 waves given SRS of $s_0$ with $|s_0| = 2$. Since the diameter of the population graph $G$ is 6 here, a large part of it may be observed under 4SBS, as in the case of $s_0 = \{3, 12\}$ above; indeed, $G$ is fully observed for 215 out of 780 possible initial samples. In addition, we consider induced OP following SRS of $s$,

Table 5.4 Applying BIGS-IWE for $C_4$ under $TSBS$ given $s_0 = \{3, 12\}$

| $i \in s_0$ | $\kappa \in \alpha_i^*$ | $M(\kappa)$ | $|\beta_\kappa^*|$ | $\{|\alpha_i^*| : i \in \beta_\kappa^*\}$ |
|---|---|---|---|---|
| $T = 2$ | | | | |
| 3 | A | $\{3, 8, 21, 22\}$ | 4 | $\{1, 1, 1, 1\}$ |
| 12 | B | $\{12, 13, 18, 31\}$ | 4 | $\{2, 2, 3, 1\}$ |
| $\beta_\kappa^* = M(\kappa)$ | C | $\{12, 15, 18, 32\}$ | 4 | $\{2, 1, 3, 2\}$ |
| $T = 4$ | | | | |
| 3 | A | $\{3, 8, 21, 22\}$ | 15 | — |
| 12 | B | $\{12, 13, 18, 31\}$ | 16 | — |
| $\beta_\kappa^* = M(\kappa) \cup \beta_\kappa^1(M)$ | C | $\{12, 15, 18, 32\}$ | 14 | — |
| | D | $\{13, 18, 29, 32\}$ | 12 | — |

for which $s_{\text{ref}} = s \times s$, where $|s|$ is set to be the expected number of observed nodes under $T = 1$ and $T = 2$, which are 9 and 21, respectively. Denote by $\hat{\theta}$ the resulting HTE.

Table 5.5 gives the mean squared errors (or variances) of the different estimators. The strategy BIGS-IWE is applied for $T$SBS. In case an estimator is not feasible for a certain motif under 4SBS, the result would be unavailable in the table.

Induced OP is understandably much less efficient than incident OP, as the order of the motif increases; compare e.g. the results for SRS of size 21 and 2SBS using $\beta_{\kappa}^{*} = M(\kappa)$, where both have the same expected number of nodes in the sample graph.

Under $T$SBS from Figure 5.5, the HTE $\hat{\theta}_y$ and the multiplicity estimator $\hat{\theta}_{z\beta}$ are about equally efficient for these motifs. The HH-type estimator $\hat{\theta}_{z\alpha 1}$ using the PIDA-weights can be much more efficient, especially for the lower-order motifs $\mathcal{K}_2$ and $\mathcal{S}_2$.

## BIBLIOGRAPHIC NOTES

Goodman (1961) considers first snowball sampling from a special digraph, where each node has one and only one out-edge. Frank (1977a) and Frank and Snijders (1994) consider one-wave SBS from arbitrary population graphs.

Zhang and Patone (2017) derive the HT-estimator for $T$SBS, including the illustrations in Section 5.1.

The incident OP includes all the successors $\alpha_i$. Instead, taking randomly one successor at each wave would yield a random walk (e.g. Klovdahl, 1989; Snijders, 1992). The sample inclusion probabilities are intractable under random walk sampling. A different basis of inference is needed, as will be discussed in Chapter 6.

Frank (1971) defines the *reach* at a node as the order of the connected component containing it. Provided one is able to observe the reach of a node without actually moving away from it, the ancestry knowledge would be readily available. Whether such an OP is applicable depends on the conditions one operates in.

The Lemmas 5.1 - 5.5, as well as the numerical illustration in Section 5.3, are developed by Zhang and Oguz-Alper (2020).

**Table 5.5** Mean squared errors of graph total estimators under induced OP from SRS of size 9 or 21, or SBS of maximum 4 waves given initial SRS of size 2

| Estimator | $\mathcal{K}_2$ | $\mathcal{S}_2$ | $\mathcal{K}_3$ | $\mathcal{K}_4$ | $\mathcal{C}_4$ | $\mathcal{S}_3$ | $\mathcal{P}_3$ |
|---|---|---|---|---|---|---|---|
| Induced OP, $|s| = 9$   $\hat{\theta}$ | 1 263 | 47 134 | 2 869 | 2 167 | 5 168 | 231 805 | 797 578 |
| Induced OP, $|s| = 21$   $\hat{\theta}$ | 152 | 4 533 | 198 | 41 | 116 | 11 523 | 52 488 |
| $\beta^*_\kappa = M(\kappa)$   $\hat{\theta}_y$ | 471 | 5 269 | 193 | 10 | 38 | 5 092 | 27 717 |
| $\hat{\theta}_{z\beta}$ | 475 | 5 447 | 199 | 10 | 39 | 5 368 | 29 441 |
| $\hat{\theta}_{z\alpha 1}$ | 116 | 613 | 160 | 10 | 28 | – | – |
| $\beta^*_\kappa = M(\kappa) \cup \beta^1_\kappa(M)$   $\hat{\theta}_y$ | 306 | 1 614 | 92 | 4 | 7 | 1 382 | – |
| $\hat{\theta}_{z\beta}$ | 281 | 1 485 | 98 | 5 | 7 | 1 403 | – |

# CHAPTER 6

# Targeted random walk sampling

Walk sampling is a depth-first graph sampling method, where the initial node does not need to have a known selection probability. It can be attractive for large and often dynamic graphs if the walk is fast-moving. The stationary successive sampling probability when the walk is at equilibrium can provide the basis of inference. We consider estimation by walk sampling of graph parameters defined for finite-order induced motifs.

## 6.1  RANDOM WALK IN GRAPHS

### 6.1.1  Basics

Let $G = (U, A)$ be a simple graph. Let $X_t = i$ be the node (or *state*) at step time $t$. Let $a_{i+}$ be the number of out-edges from node $i$. For $t + 1$, one selects one of the out-edges randomly, $(ij) \in A_{i+}$, which yields $X_{t+1} = j$ as the next state of the random walk. Thus, $\{X_0, X_1, X_2, ...\}$ form a Markov chain, where $X_0$ is the initial state and the transition probability is

$$p_{ij} := \Pr(X_{t+1} = j | X_t = i) = \frac{a_{ij}}{a_{i+}}$$

Let $P$ be the $N \times N$ matrix of *transition probabilities* with the elements $p_{ij}$. Let $p_0$ be the *row* $N$-vector of initial node selection probabilities, the probabilities of $X_t$ are given by

$$p_t = p_0 P^t$$

A random walk reaches gradually its *equilibrium*, if the chance that it visits a given node depends less and less on the initial state $X_0$. The *stationary probability* of $X_t = i$ is the fraction of times node $i$ is visited when the walk is at equilibrium, denoted by

$$\pi_i = \Pr(X_t = i) \quad \text{and} \quad \pi = (\pi_1, ..., \pi_N)$$

where $\sum_{i \in U} \pi_i = 1$. If $G$ consists of a single component where every node can be reached from any other in time, the chain $\{X_t\}$ is irreducible. We have

$$\pi_j = \sum_{i \in U} \pi_i p_{ij} \quad \text{or} \quad \pi = \pi P \tag{6.1}$$

where $\pi$ is the left eigenvector of $P$ with eigenvalue 1. For undirected graphs, where $a_{i+} = d_i$ is the degree of node $i$, we have

$$\pi_i = \frac{d_i}{2R} \tag{6.2}$$

where $R = |A|$. Moreover, for two adjacent nodes $i$ and $j$, the *flow probability* is the same in either direction at equilibrium:

$$\pi_i p_{ij} = \pi_j p_{ji} \tag{6.3}$$

The stationary probabilities $\pi$ defined by (6.1) cannot be given explicitly for digraphs.

Only by walking one cannot move beyond the component, to which the initial node belongs. Random jumps can be introduced. At each step time, if possible let the walk move to an adjacent node with probability $r$ or jump randomly to *any* node in $U$ with probability $1 - r$. The transition probabilities are then given by

$$p_{ij} = \begin{cases} r\frac{a_{ij}}{a_{i+}} + (1-r)\frac{1}{N} & \text{if } a_{i+} > 0 \\ \frac{1}{N} & \text{if } a_{i+} = 0 \end{cases} \tag{6.4}$$

We have

$$p_{t+1} = r p_t P + (1-r)u$$

in terms of the $N$-vector

$$u = \left(\frac{1}{N}, \cdots, \frac{1}{N}\right)$$

Adding random jumps ensures the chain $\{X_t\}$ is irreducible. Since it is possible to return to the same node at any step time, the chain is aperiodic. The limiting probabilities are the stationary probabilities, which are given by

$$\pi = (1-r)u(I - rP)^{-1}$$

where $I$ is the identity matrix. The Taylor expansion is given by

$$\pi_i \approx \frac{1}{N} + \sum_{l=1}^{\infty} \frac{r^l}{N} \sum_{j=1}^{N} \left\{ \left(\frac{a_{ji}}{a_{j+}}\right)^l - \left(\frac{a_{ji}}{a_{j+}}\right)^{l-1} \right\}$$

### 6.1.2 Examples of related topics

*PageRank*   The random-walk stationary probability $\pi_i$ is a graph centrality measure of the node. One can allow $u_i$ to differ for the nodes in $U$, in which case $u$ is referred to as the preference vector. The probabilities $\pi$ can e.g. be used to rank web pages, where it is known as PageRank and $r$ in (6.4) is called the "damping factor" (Brin and Page, 1998). Given $u_i = a_{+i}/\sum_{j \in U} a_{+j}$, the Taylor expansion becomes

$$\pi_i \approx \frac{a_{+i}}{\sum_{j \in U} a_{+j}} + \sum_{l=1}^{\infty} \frac{r^l}{\sum_{j \in U} a_{+j}} \sum_{j=1}^{N} (a_{+j} - a_{j+}) \left(\frac{a_{ji}}{a_{j+}}\right)^l$$

Given $r > 0$, page $i$ receives a positive contribution from page $j$ that has a larger in-degree than out-degree, $a_{+j} > a_{j+}$, whereas the contribution is negative in the opposite case.

*Betweenness*   The shortest-path (SP) *betweenness* of a node $i$ is the fraction of shortest paths between the pairs of nodes in a graph which pass through $i$. In cases of more than one shortest path between a given pair of nodes, each of them is given an equal weight such that the weights sum to one. The denominator of the fraction is $N(N-1)/2$ for undirected graphs.

In Figure 6.1, the SP betweenness of ★ is 0, because it is always "short-circuited" by the two ○ nodes. Thus, a node may be never on a shortest path, although it may seem intuitively important to the flows in a graph.

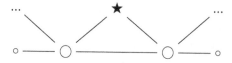

Figure 6.1   An illustration for betweenness

Roughly speaking, the random-walk (RW) betweenness of a node is related to the number of times that a random walk between a pair of nodes passes through it, averaged over all the node pairs in a graph. In Figure 6.1, a random walk through either ◯ has the same chance of moving to ★ or the other ◯. Thus, the RW betweenness of ★ is much higher than its SP betweenness.

*Core-periphery structure*   One may wish to decompose the nodes of a connected graph into one or several densely-connected cores along with sparsely-connected peripheral nodes. The nodes in a core are heavily interconnected among themselves, whereas the peripheral nodes are predominantly adjacent to the core nodes but sparsely connected among themselves. Consequently, a random walk currently located at a peripheral node is much less likely to move to another peripheral node at the next step time. Let the persistent probability of a set of nodes $M$ be given by

$$\xi_M = \sum_{i,j \in M} \pi_i p_{ij} / \sum_{i \in M} \pi_i$$

The idea is that the nodes in $M$ tend to be peripheral if $\xi_M$ is small. One can start with $M$ consisting of a node that has the lowest total in and out-degree and one-by-one add nodes to $M$, such that each time the increment of $\xi_M$ is the minimum amount possible. The nodes entering later into $M$ are considered deeper into a core of the graph.

*Respondent driven sampling (RDS)*   One may lack direct access to a target group of individuals, which is small in size. To obtain a reasonable number of observations from the group of interest, the only practical method may be to follow the links between its members, even though such links are not necessarily exclusive to them. A peculiar feature of RDS is that link-tracing takes place when an in-sample individual (current state) recruits one of its neighbours (adjacent nodes), and participation is usually rewarded financially. The out-degree of each node needs to be collected, in order to *model* RDS as a random walk. Taking a random jump amounts to selecting a new initial node if the last recruit fails to supply any individual.

## 6.2 TARGETED RANDOM WALK

A walk may be said to be *targeted* if its stationary probability $\pi_i$ is subject to one's choice beyond that of a pure random walk.

### 6.2.1 Metropolis-Hastings (MH) method

Now that the flow probability at equilibrium (6.3) is equal in either direction, one can apply an adjustment to the pure random-walk transition probabilities, in order to achieve the *uniform* stationary probability by a walk, where

$$\pi_i \equiv 1/N$$

Take two adjacent nodes $i$ and $j$. If $a_{i+} < a_{j+}$, then to increase the fraction of time spent on $i$, compared to that under the random walk, one can introduce a probability of not moving from $i$ to $j$, when $j$ is initially selected by the random walk, while always accepting a move from $j$ to $i$ whenever $i$ is selected by the random walk. Clearly, to achieve $\pi_i = \pi_j$ given $a_{i+} < a_{j+}$, the probability of accepting a move from $i$ to $j$ should be $a_{i+}/a_{j+}$. The same holds also when the walk includes random jumps.

Adding such an acceptance-rejection mechanism to a Markov chain is called the Metropolis-Hastings (MH) method. Let the right-hand side of (6.4) be the *proposal probabilities*, denoted by $q_{ij}$, so that $p_{ij}$ does still denote the transition probability. Let the *acceptance probability* for a move from $i$ to $j$ be

$$\psi_{ij} = \min\left\{\frac{q_{ji}}{q_{ij}}, 1\right\}$$

Let the transition probabilities from $i$ be given by

$$\begin{cases} p_{ij} = q_{ij}\psi_{ij} & \text{if } i \neq j \\ p_{ii} = 1 - \sum_{j \neq i} p_{ij} \end{cases}$$

As explained above, these transition probabilities would result in the uniform stationary probability. Such a walk is referred to as the *uniform walk*.

Table 6.1 provides the details of the acceptance probabilities $\psi_{ij}$ in different situations. We have $\psi_{ij} = 1$ in the third row and $\psi_{ij} < 1$ in the third last row, because $a_{i+} \leq N$. If $N \gg \max(a_{i+}, a_{j+})$ in the second last row, then

$$\psi_{ij} \approx \frac{a_{i+}}{a_{j+}}$$

which is the modification needed to achieve a uniform walk in connected graphs. Every situation is possible in directed graphs, whereas the last column indicates whether or not a situation arises in undirected graphs.

To obtain a targeted walk with the other specified stationary probabilities $\pi$, one can use the acceptance-rejection mechanism given by

$$\psi_{ij} = \min\left\{\frac{\pi_j q_{ji}}{\pi_i q_{ij}}, 1\right\} \qquad (6.5)$$

to be referred to as the *MH-targeted walks*. In particular, notice that one needs to observe all the successors of all the successors of the current node $i$, i.e. $\alpha_j$ for all $j \in \alpha_i$, before one can make a move, since $q_{ji}$ is needed in (6.5).

One can set $\pi_i \propto 1 + a_{i+}$, resulting in the *degree+1* walk, which accommodates the nodes with 0 out-degree. For a pair of adjacent nodes with $a_{i+} < a_{j+}$, where $N \gg \max(a_{i+}, a_{j+})$, the acceptance probability is then given by

$$\psi_{ij} = \frac{(1-r)(a_{j+}+1)/N + r(1+1/a_{j+})}{(1-r)(a_{i+}+1)/N + r(1+1/a_{i+})} \approx \frac{1+1/a_{j+}}{1+1/a_{i+}}$$

One can set $\pi_i \propto \pi(y_i)$ as a function of the $y$-value associated with each node, which yields an adaptive sampling method. For instance, suppose $y_i$ is binary, where one is interested in the nodes with $y_i = 1$ but not those with $y_i = 0$. Setting $\pi(1) = 2$ and $\pi(0) = 1$ means that the former is tuned to have a stationary probability twice that of the latter.

## 6.2.2 Targeted random walk

Let an imaginary node, denoted by $\star \notin U$, be connected to all the nodes in an undirected graph $G$, such that a random jump can

**Table 6.1** Acceptance probabilities $\psi_{ij}$ for $i \neq j$ ( ✓ if relevant in undirected graphs)

| Situation | | $q_{ji}/q_{ij}$ | $\psi_{ij}$ | Undirected $G$ |
|---|---|---|---|---|
| $a_{i+}=0$ | $a_{j+}=0$ | $N/N=1$ | $1$ | ✓ |
| $a_{i+}=0$ | $a_{ji}=0$ | $N(1-r)/N$ | $1-r$ | ✓ |
| $a_{j+}>0$ | $a_{ji}=1$ | $(1-r)+rN/a_{j+}$ | $1$ | – |
| $a_{i+}>0$ | $a_{ij}=0$ | $N/(1-r)N$ | $1$ | ✓ |
| $a_{j+}=0$ | $a_{ij}=1$ | $\dfrac{1/N}{(1-r)/N+r/a_{i+}}$ | $\dfrac{1}{1+r(N/a_{i+}-1)}$ | – |
| $a_{i+}>0$, $a_{j+}>0$ | $a_{ij}=0,\,a_{ji}=0$ | $N(1-r)/(1-r)N$ | $1$ | ✓ |
| | $a_{ij}=0,\,a_{ji}=1$ | $\dfrac{(1-r)/N+r/a_{j+}}{(1-r)/N}$ | $1$ | – |
| | $a_{ij}=1,\,a_{ji}=0$ | $\dfrac{(1-r)/N}{(1-r)/N+r/a_{i+}}$ | $\dfrac{1-r}{1+r(N/a_{i+}-1)}$ | – |
| | $a_{ij}=a_{ji}=1$ $a_{i+}<a_{j+}$ | $\dfrac{(1-r)/N+r/a_{j+}}{(1-r)/N+r/a_{i+}}$ | $\dfrac{(1-r)+rN/a_{j+}}{(1-r)+rN/a_{i+}}$ | ✓ |
| | $a_{ij}=a_{ji}=1$ $a_{i+}>a_{j+}$ | $\dfrac{(1-r)/N+r/a_{j+}}{(1-r)/N+r/a_{i+}}$ | $1$ | ✓ |

be given as two successive adjacent moves via $\star$. Let $X_t = i$ at time step $t$. Let the probability of moving to $\star$ be $r_i = r/(d_i + r)$ and, having moved to $\star$, one takes immediately another random move away from it to reach $X_{t+1} = j$, for some $j \in U$. Notice that the probability of taking a random jump $r_i$ is not constant across the nodes. The procedure only requires one to observe $\alpha_i$ before making the next move.

The transition probability from $X_t = i$ to $X_{t+1} = j$ is now given by

$$p_{ij} = \begin{cases} \frac{1}{d_i+r}\left(1 + \frac{r}{N}\right) & \text{if } a_{ij} = 1 \\ \frac{r}{d_i+r}\left(\frac{1}{N}\right) & \text{if } a_{ij} = 0 \text{ including } i = j \end{cases} \qquad (6.6)$$

Since

$$d_j + r = \sum_{i \in \alpha_j} \frac{d_i + r}{d_i + r}\left(1 + \frac{r}{N}\right) + \sum_{i \notin \alpha_j} \frac{d_i + r}{d_i + r}\left(\frac{r}{N}\right)$$

the resulting stationary probability in undirected graphs is

$$\pi_i \propto d_i + r \qquad (6.7)$$

We shall refer to (6.6) as the *targeted random walk (TRW)*. It has the same stationary probability as the degree+1 walk if $r = 1$, without applying the MH method. It is close to pure random walk, if $r$ is a small positive constant, but without the latter's difficulty in graphs with multiple components.

### 6.2.3 Generalised ratio estimator for 1st-order parameter

Given any constants $\{y_i : i \in U\}$ associated with the nodes of the graph, Let $\mu = \theta/N$ be a 1st-order graph parameter, where $\theta = \sum_{i \in U} y_i$ and $N = |U|$.

Let a targeted walk have stationary probabilities $\pi_i \propto c_i$, where $c_i$ may be unknown for the unvisited nodes. Uniform walk is the special case with $c_i \equiv 1$. Random walk in an undirected connected graph is the special case with $c_i = d_i$. TRW is the special case with $c_i = d_i + r$ in undirected graphs.

A walk is stationary *draw-by-draw* at equilibrium, now that $\pi$ is the same for each draw. To estimate $\mu$, one can use an extraction of $n$ states, which need not to be successive, denoted by

$$\mathfrak{s}_n = \{X_{t_1}, X_{t_2}, ..., X_{t_n}\} \quad \text{with} \quad t_1 < t_2 < \cdots < t_n$$

As a convention to allow for $t_n = T$, where $T$ is the last time step, we assume that the OP is applied to $X_T$ generally. A *generalised ratio estimator* of $\mu$ is then given by

$$\hat{\mu} = \left(\frac{1}{n}\sum_{i\in\mathfrak{s}_n}\frac{y_i}{c_i}\right)\bigg/\left(\frac{1}{n}\sum_{i\in\mathfrak{s}_n}\frac{1}{c_i}\right) = \sum_{i\in\mathfrak{s}_n}\frac{y_i}{c_i}\bigg/\sum_{i\in\mathfrak{s}_n}\frac{1}{c_i} \qquad (6.8)$$

It is *approximately* unbiased for $\mu$ given sufficiently large $n$.

One can reduce the within-walk auto-correlations among the states in $\mathfrak{s}_n$ by extracting time steps that are far apart from each other, in order to treat $\mathfrak{s}_n$ approximately as an IID sample for variance estimation. An alternative is to administer multiple walks independently, whereby one can average the estimators from all the walks and use the between-walk variance as the basis for variance estimation, regardless the within-walk auto-correlations.

## 6.3   STRATEGY FOR TRW SAMPLING

Using a targeted walk and the estimator (6.8) is a strategy based on the draw-by-draw stationary probability not the sample inclusion probability. The initial node, as a singleton sample $s_0 = \{X_0\}$, does not need to have a known selection probability. However, one can only deal with 1st-order graph parameters in this way. A more general approach is needed to finite-order graph parameters by *targeted random walk sampling (TRWS)* from undirected graphs, or by other targeted walk sampling methods.

### 6.3.1   Sample graph

Consider $T$-step targeted random walk sampling ($T$TRWS). The definition of sample graph $G_s$ by (1.3) accommodates any isolated node in $G$, which is visited by random jumps but is not incident to $A_s$ even though the walk passes through it. Now that the OP is applied to $X_T$ generally, the seed sample is

$$s = \{X_0, X_1, ..., X_T\}$$

Notice that a node can appear more than once in the seed sample of walk sampling, whereas by definition the seed sample nodes are all distinct under $T$SBS. Under $T$TRWS, we observe $\{i\} \times \alpha_i$ and $\alpha_i \times \{i\}$ for any $i \in s$, such that the reference set is

$$s_{\text{ref}} = s \times U \cup U \times s$$

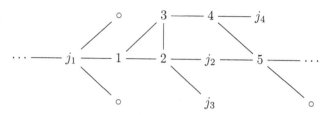

**Figure 6.2** An illustration of walk in graph

There are many observed motifs by a walk beyond the nodes visited. Take Figure 6.2 for illustration, which passes ... 1, 2, 3, 4, 5... in succession. Below are some motifs observed by TRW.

- $[\{1, 2\}]$ is observed at $X_t = 1$, where $2 \in \alpha_1$.

- $[\{1, 2, 3\}]$ is observed based on $(X_t, X_{t+1}) = (1, 2)$, where we observe $3 \in \alpha_1$ at $X_t = 1$.

- $[\{1, 2, 3, 4\}]$ is observed based on $(X_t, X_{t+1}, X_{t+2}) = (1, 2, 3)$, where we observe $4 \notin \alpha_1$ at $X_t = 1$ and $4 \notin \alpha_2$ at $X_{t+1} = 2$.

- $[\{1, 3, 5\}]$ is observed based on $(X_t, X_{t+1}, X_{t+2}) = (1, 2, 3)$, where we observe $5 \notin \alpha_1$ at $X_t = 1$.

An induced motif $[M]$ is observed if $M \times M \subseteq s_{\text{ref}}$, either when $M \subseteq s$ or if only one node in $M$ falls outside of $s$.

Suppose $X_0 = 1$, $X_T = 4$ and $s = \{1, 2, 3, 4\}$ in Figure 6.2. Since the nodes in $s$ are all distinct here, the number of observed motifs in $s \times s$ is $2^4 - 1 = 15$. In addition, the nodes $(\alpha(s) \cup \beta(s)) \setminus s$ are observed, as are the edges in $s \times (\alpha(s) \setminus s) \cup (\beta(s) \setminus s) \times s$, such that there are many observed motifs in addition to those in $s \times s$.

*MH-targeted walk*    The reference set of $T$-step MH-targeted walk is $s_{\text{ref}} = s \times U \cup U \times s$, where

$$s = \{X_0, \alpha_{X_0}, X_1, \alpha_{X_1}, ..., X_T, \alpha_{X_T}\}$$

Let $X_0 = 1$ and $X_T = 4$ in Figure 6.2. We have

$$s = \{1, j_1, 2, 3, j_2, j_3, 4, j_4, 5\}$$

after removing the duplicates. The number of observed motifs in $s \times s$ is then $2^9 - 1 = 511$.

## 6.3.2 Stationary successive sampling probability

Let $M = \{X_{t_1}, ..., X_{t_q}\}$ be a set of states by TRWS, where $q = |M|$ and $t_1 < \cdots < t_q$. At equilibrium, we have

$$\pi_M = \Pr(X_{t_1}, ..., X_{t_q}) = \pi_{X_{t_1}} \prod_{i=1}^{q-1} p(X_{t_i}, X_{t_{i+1}}) \qquad (6.9)$$

where $p(X_{t_i}, X_{t_{i+1}})$ is the transition probability from $X_{t_i}$ to $X_{t_{i+1}}$ over exactly $t_{i+1} - t_i$ steps, which is given by the $(X_{t_i}, X_{t_{i+1}})$-th element of the matrix $P^{t_{i+1}-t_i}$.

The expression (6.9) includes $\pi_i$ as the 1st-order special case of $|M| = 1$. Suppose $(X_t, X_{t+2}, X_{t+4}) = (1, 3, 5)$, we have

$$\pi_{X_t X_{t+2} X_{t+4}} = \pi_1 \left( \sum_{i \in U} p_{1i} p_{i3} \right) \left( \sum_{i \in U} p_{3i} p_{i5} \right)$$

which requires the entire 1st and 3rd rows and 3rd and 5th columns of the matrix $P$. Suppose $(X_t, X_{t+1}, X_{t+2}, X_{t+3}) = (1, 2, 3, 4)$, we have

$$\pi_{X_t X_{t+1} X_{t+2} X_{t+3}} = \pi_1 p_{12} p_{23} p_{34}$$

which requires only the transition probabilities of successive states.

We shall refer to $\pi_M$ given by (6.9) as the *stationary successive sampling probability (S3P)*, when $M = \{X_{t+1}, ..., X_{t+q}\}$ is a *sequence* of successive states by walk sampling.

Apart from the actual sequences in a TRW, the S3P (6.9) is also known up to a proportionality constant for any *hypothetical* sequence $M$, given $M \subseteq s$, because the sub-matrix of $P$ corresponding to $s \times s$ is known even when the full matrix $P$ is unavailable. For instance, given the actual sequence $(X_t, X_{t+1}, X_{t+2}) = (1, 2, 3)$, we can calculate the transition probability $p_{32} p_{21}$ of a hypothetical sequence $(X_t, X_{t+1}, X_{t+2}) = (3, 2, 1)$ as well.

Let the *generating (sequences of) states* of $T$TRWS with seed sample $s$ be

$$\mathcal{C}_s = \{M : M \subseteq s\} \qquad (6.10)$$

The subset containing the parts of the actual walk is given by

$$\mathcal{C}_w = \{\{X_t, ..., X_{t+q}\} : 0 \le t \le t + q \le T\} \qquad (6.11)$$

Suppose TRW with $X_0 = 1$ and $X_T = 4$ in Figure 6.2, where

$s = \{1, 2, 3, 4\}$. We have

$$\mathcal{C}_w = \{\{1\}, \{1, 2\}, \{1, 2, 3\}, \{1, 2, 3, 4\},$$
$$\{2\}, \{2, 3\}, \{2, 3, 4\}, \{3\}, \{3, 4\}, \{4\}\}$$

Apart from $\mathcal{C}_w$, the S3P is also known proportionally for the other elements in $\mathcal{C}_s$, such as $M = \{4, 2, 1\}$ with $\pi_M = \pi_4 p_{42} p_{21}$.

### 6.3.3 Eligible sample motifs

Consider the triangle motif $\kappa$ with $M = \{1, 2, 3\}$ in Figure 6.2 under $TTRWS$. It is actually observed from $(X_t, X_{t+1}) = (1, 2)$ with S3P $\pi_1 p_{12}$. Let

$$\tau_\kappa = \sum_{(i,j) \in U} \frac{\delta_{i,j}}{\pi_{i,j}} I_\kappa(i, j)$$

where $\delta_{i,j} = 1$ if $(X_t, X_{t+1}) = (i, j)$ and 0 otherwise, $\pi_{i,j} = \pi_i p_{ij} = E(\delta_{i,j})$, and $I_\kappa(i, j) = 1$ if $\kappa$ is observed given $(X_t, X_{t+1}) = (i, j)$ and 0 otherwise. We have

$$E(\tau_\kappa) = \sum_{(i,j) \in U} I_\kappa(i, j) = 6$$

since $I_\kappa(i, j) = 1$ if $(X_t, X_{t+1}) = $ (1,2), (2,1), (1,3), (3,1), (2,3) or (3,2). One needs to take into account all these 6 possibilities, when estimating a graph total over $\Omega$ that includes $\kappa$ if one of them is the realised $(X_t, X_{t+1})$.

Note that a motif can be observed, for which it is infeasible to take such considerations. For example, suppose the triangle motif $\kappa$ with $M = \{1, 2, 3\}$ is observed given $(X_t, X_{t+1}, X_{t+2}) = (1, i, 2)$ for $i \notin \{1, 2, 3\}$. One cannot account for all such 3-state sequences as long as $U \setminus s \neq \emptyset$.

Let the sample motif $\kappa$ be observed from the *actual sampling sequence of states (AS3)* $s_\kappa = (X_t, ..., X_{t+q})$, for some time step $t$ and $q = |s_\kappa| - 1$. An *equivalent sampling sequence of states (ES3)* of $s_\kappa$, denoted by $\tilde{s}_\kappa \sim s_\kappa$, is any possible sequence of states of the same length $|\tilde{s}_\kappa| = |s_\kappa|$, such that the motif $\kappa$ would be observed given $(X_t, X_{t+1}, ..., X_{t+q}) = \tilde{s}_\kappa$ but not based on any subsequence of $\tilde{s}_\kappa$. In particular, we have $s_\kappa \sim s_\kappa$.

The AS3 of the motif of $M = \{1, 2, 3\}$ is $(X_t, X_{t+1}) = (1,2)$ in Figure 6.2, and the other ES3 are $(X_t, X_{t+1}) = (2,1)$, $(1,3)$, $(3,1)$, $(2,3)$ and $(3,2)$. The AS3 of the motif of $M = \{1, 2, 4\}$ is also $(X_t, X_{t+1}) = (1,2)$, where the other ES3 are $(X_t, X_{t+1}) = (2,1)$, $(1,4)$, $(4,1)$, $(2,4)$ or $(4,2)$.

Take $M = \{2, 4, 5\}$, its AS3 is $(X_t, X_{t+1}, X_{t+2}) = (2, 3, 4)$. Any $(X_t, X_{t+1}, X_{t+2}) = (2, i, 4)$ is its ES3, where $i \notin \{4, 5\}$, but not $(2, 4, 4)$, because $[M]$ would be observed given the subsequence $(X_t, X_{t+1}) = (2, 4)$; similarly for $(2, 5, 4)$. The other ES3s are $(2, i, 5)$ where $i \notin \{4, 5\}$, $(4, i, 2)$ and $(4, i, 5)$ where $i \notin \{2, 5\}$, $(5, i, 2)$ and $(5, i, 4)$ where $i \notin \{2, 4\}$.

Under TRWS at equilibrium, a motif $\kappa$ observed from AS3 $s_\kappa$ can be sampled *sequence-by-sequence*, for which its ES3s constitute its multiplicity, denoted by

$$F_\kappa = \{\tilde{s}_\kappa : \tilde{s}_\kappa \sim s_\kappa\} \tag{6.12}$$

where the S3P $\pi_{\tilde{s}_\kappa}$ is known proportionally if $\tilde{s}_\kappa \in \mathcal{C}_s$.

**Lemma 6.1.** *Under TRWS at equilibrium, a sample motif $\kappa \in \Omega_s$ observed from AS3 $s_\kappa$ is eligible for estimation if $F_\kappa \subseteq \mathcal{C}_s$, where $\mathcal{C}_s$ is given by (6.10).*

To illustrate, let $s = \{1, 2, 3, 4\}$ in Figure 6.2. The motif $[\{j_1, 1\}]$ has $s_\kappa = \{1\}$ and $F_\kappa = \{1, j_1\}$, such that it is ineligible since $j_1 \notin s$. The motif $[\{1, 2, 3\}]$ has AS3 $s_\kappa = \{1, 2\}$ from $(X_1, X_2) = (1, 2)$, as well as $s_\kappa = \{2, 3\}$ from $(X_2, X_3) = (2, 3)$. It can be used twice for estimating a graph total of triangles, with the same

$$F_\kappa = \{(1, 2), (2, 1), (1, 3), (3, 1), (2, 3), (3, 2)\}$$

Both $\kappa'$ of $\{1, 2, 4\}$ and $\kappa''$ of $\{1, 2, 5\}$ have the same AS3 $(X_1, X_2) = \{1, 2\}$. The motif $\kappa'$ is eligible since $\{1, 2, 4\} \subset s$, whereas $\kappa''$ is ineligible since $5 \notin s$ and

$$F_{\kappa''} = \{(1, 2), (2, 1), (1, 5), (5, 1), (2, 5), (5, 2)\} \not\subset \mathcal{C}_s$$

### 6.3.4  Generalised ratio estimator

Estimation of finite-order graph parameters from TRWS sequence-by-sequence generalises estimation of 1st-order parameters from TRWS draw-by-draw.

Given any TRWS from $G$, choose a seed sample of successive states at equilibrium, which is of size $n$, after the initial burn-in states, denoted by

$$s = \{X_1, ..., X_n\}$$

Obtain the motif sample $\Omega_s$, the generating states $\mathcal{C}_s$, and its subset $\mathcal{C}_w$. For any $\kappa \in \Omega_s$, let $s_\kappa$ be its AS3, where $s_\kappa \in \mathcal{C}_w$, and let $F_\kappa$ contain all its ES3s.

Let $\theta = \sum_{\kappa \in \Omega} y_\kappa$ be a graph total in terms of (1.2). Let sample motif $\kappa \in \Omega_s$ have AS3 $s_\kappa = (X_t, ..., X_{t+q})$ with known $F_\kappa$. Let $M$ be a possible sequence of states $(X_t, ..., X_{t+q})$; let the sampling indicator $\delta_M = 1$ if $M$ is realised and 0 otherwise, with the associated S3P $\pi_M$. Let the observation indicator $I_\kappa(M) = 1$ if $M \in F_\kappa$ and 0 otherwise. Let $w_{M\kappa}$ be the *incidence weight* for any $M \in F_\kappa$, let $w_{M\kappa} = 0$ if $M \notin F_\kappa$, such that for any $\kappa \in \Omega$, we have

$$\sum_{M \in F_\kappa} w_{M\kappa} = \sum_M I_\kappa(M) w_{M\kappa} = 1$$

The corresponding IWE is given by

$$\hat{\theta}(X_t, ..., X_{t+q}) = \sum_{\kappa \in \Omega} \sum_M \frac{\delta_M}{\pi_M} I_\kappa(M) w_{M\kappa} y_\kappa \qquad (6.13)$$

which is unbiased for $\theta$, since

$$E\big(\hat{\theta}(X_t, ..., X_{t+q})\big) = \sum_{\kappa \in \Omega} y_\kappa \sum_M I_\kappa(M) w_{M\kappa} = \sum_{\kappa \in \Omega} y_\kappa$$

That is, one can consider $F_\kappa$ as the ancestors of $\kappa$ over repeated sampling of $(X_t, ..., X_{t+q})$, and apply the strategy BIGS-IWE with $\beta_\kappa = F_\kappa$ in (3.2).

Although there can be infinitely many possibilities, the two basic choices of $w_{M\kappa}$ are the multiplicity weight given by

$$w_{M\kappa} = 1/|F_\kappa| \qquad (6.14)$$

for any $M$ in $F_\kappa$, and the *proportional-to-probability weight (PPW)* in case of eligible sample motifs given by

$$w_{M\kappa} = \pi_M / \pi_{(\kappa)} \qquad \text{and} \qquad \pi_{(\kappa)} = \sum_{M \in F_\kappa} \pi_M \qquad (6.15)$$

Given the AS3 $s_\kappa$ is of the order $|s_\kappa| = q + 1$ and $|s| = n$, there are at most $n - q$ possible estimators based on $(X_t, ..., X_{t+q})$ and given by (6.13), denoted by $\hat{\theta}_t$ for $t = 1, ..., n - q$. Let $\mathbb{I}_t = 1$ if $(X_t, ..., X_{t+q})$ can lead to the observation of at least one motif in $\Omega$, in which case $\hat{\theta}_t$ exists, and 0 otherwise. Provided TRWS is stationary sequence-by-sequence, an estimator of $\theta$ combining all the $\hat{\theta}_t$ is given by

$$\hat{\theta} = \Big( \sum_{t=1}^{n-q} \mathbb{I}_t \hat{\theta}_t \Big) / \Big( \sum_{t=1}^{n-q} \mathbb{I}_t \Big). \tag{6.16}$$

Insofar as $\pi_M$ in (6.13) contains an unknown proportionality constant, (6.16) cannot be calculated. However, any function of graph totals that is invariant towards the unknown proportionality constant, when each total is replaced by its estimator (6.16), can be estimated using a generalised ratio estimator similar to (6.8).

For instance, let $\mu = \theta/N_\Omega$, where $N_\Omega = |\Omega|$. A generalised ratio estimator is given by $\hat{\mu} = \hat{\theta}/\hat{N}_\Omega$, where $\hat{N}_\Omega$ is given by (6.13) and (6.16) with $y_\kappa \equiv 1$. This reduces to (6.8) in the special case of $\Omega = U$, $N_\Omega = N = |U|$ and $q = 0$. Or, let

$$\mu = \theta/\theta' \quad \text{where} \quad \theta = \sum_{\kappa \in \Omega} y_\kappa \quad \text{and} \quad \theta' = \sum_{\kappa \in \Omega'} y'_\kappa$$

refer to different kinds of motif in $\Omega$ and $\Omega'$, respectively. Replacing $\theta$ by $\hat{\theta}$ and $\theta'$ by $\hat{\theta}'$, each using (6.16), we obtain a generalised ratio estimator $\hat{\mu}$ of $\mu$, as long as $\hat{\mu}$ does not depend on the unknown proportionality constant in the relevant S3Ps.

## 6.4 ILLUSTRATIONS

Let $G = (U, A)$ be an undirected simple graph with 100 nodes, $N = |U| = 100$. Let $y = 1$ be the value associated with the first 20 nodes $i = 1, ..., 20$, or the cases; and let $y = 0$ be the value for the rest 80 nodes, or the noncases.

The edges are generated randomly, with different probabilities for a pair of nodes: if both have $y = 1$; if one of them has $y = 1$ and the other $y = 0$; if both have $y = 0$. In the resulting graph, there are 299 edges, $|A| = 299$; the cases have an average degree 13.5, and the noncases have an average degree 4.1. The population graph $G$ exhibits a mild core-periphery structure.

This valued graph will be held fixed for the illustrations below.

## 6.4.1 Convergence to equilibrium

Let $p_{t,i} = \Pr(X_t = i)$, for $i \in U$. We have $p_{t,i} \to \pi_i \propto d_i + r$ by (6.7) for TRW, as $t \to \infty$. How quickly a walk reaches equilibrium is affected by the selection of the initial state $X_0$. In the extreme case, where $\Pr(X_0 = i) = \pi_i$ for $i \in U$, the walk is at equilibrium from the beginning. To explore the speed of convergence, consider two other choices, where $p_{0,i} = \Pr(X_0 = i) \equiv 1/N$ or $p_{0,1} = 1$. To track the convergence empirically, we use $B$ independent simulations of the TRW to estimate

$$E(Y_t) = \sum_{i \in U} p_{t,i} y_i$$

targeted at the equilibrium expectation $E(Y_\infty) = \sum_{i \in U} \pi_i y_i$.

**Table 6.2**  $E(Y_t)$ by $t$, $r$ and initiation of $X_0$, $10^5$ simulations

| Initiation | $r = 1$, $E(Y_\infty) = 0.415$ | | | |
| | $t = 1$ | $t = 4$ | $t = 8$ | $t = 16$ |
|---|---|---|---|---|
| $p_{0,i} = \pi_i$ | 0.420 | 0.419 | 0.421 | 0.414 |
| $p_{0,i} = 1/N$ | 0.341 | 0.394 | 0.408 | 0.413 |
| $p_{0,1} = 1$ | 0.714 | 0.481 | 0.433 | 0.413 |
| Initiation | $r = 0.1$, $E(Y_\infty) = 0.447$ | | | |
| | $t = 1$ | $t = 4$ | $t = 8$ | $t = 16$ |
| $p_{0,i} = \pi_i$ | 0.449 | 0.451 | 0.450 | 0.448 |
| $p_{0,i} = 1/N$ | 0.355 | 0.412 | 0.433 | 0.449 |
| $p_{0,1} = 1$ | 0.741 | 0.529 | 0.471 | 0.448 |

Table 6.2 shows the results for $t = 1, 4, 8, 16$ and $r = 1, 0.1$, each based on $10^5$ simulations. The equilibrium expectation $E(Y_\infty)$ varies with $r$, which affects the transition probabilities. Of the two choices here, the value $r = 1$ yields the degree+1 walk, whereas the value $r = 0.1$ tunes the walk closer to pure random walk.

It can be seen that TRW stays at equilibrium if $p_{0,i} = \pi_i$. For a given value of $r$, the differences as $t$ varies provide an appreciation of the simulation error. Under the current set-up, convergence to equilibrium is apparently achieved at $t = 16$, whether the initial $X_0$ is selected completely randomly from $U$ given $p_{0,i} = 1/N$, or fixed at $i = 1$ given $p_{0,1} = 1$. Neither does the speed of convergence vary much for the values of $r$ here.

## 6.4.2 Estimation of case prevalence

To estimate the population case prevalence $\mu = \sum_{i \in U} y_i/N$, let $s = \{X_0, ..., X_T\}$ be the states obtained by $T$TRWS, where $X_0$ is drawn with $p_{0,i} = 1/N$. Apply (6.8) to $s$ yields $\hat{\mu}$. The burn-in stage is quite short in the present setting. In any case, using all the states can be instructive for appreciating the convergence of $\hat{\mu}$.

Given $T$ and $r$, generate TRW independently $B$ times, each resulting in a replicate of the estimator $\hat{\mu}$. The mean of the $B$ replicates is an estimate of $E(\hat{\mu})$ under $T$TRWS, and the variance of them is an estimate of $V(\hat{\mu})$. Calculate also the naïve variance estimate $s_T^2/(T + 1)$, where $s_T^2 = \sum_{t=0}^{T}(y_{X_t} - \bar{y})^2/T$ and $\bar{y}$ are the sample variance and mean of $\{y_{X_t} : t = 0, ..., T\}$, respectively. This is not a consistent variance estimator due to the within-walk auto-correlations. Moreover, let $\mu(1-\mu)/(T+1)$ be the theoretical variance when one can estimate $\mu$ based on an IID sample, which is drawn randomly and with replacement from $\{y_i : i \in U\}$.

Let $\psi$ be the *traverse* of the walk, given as the ratio between the number of distinct nodes visited by the walk and $N$, which indicates how extensively the walk has travelled through $G$.

Table 6.3 Estimation of case prevalence $\mu = 0.2$ under $T$TRWS

|  | $T$ | Mean($\hat{\mu}$) | SE($\hat{\mu}$) | Naïve SE | SE-IID | $\psi$ |
|---|---|---|---|---|---|---|
|  | 50 | 0.200 | 0.081 | 0.067 | 0.056 | 0.346 |
| $r = 1$ | 100 | 0.199 | 0.059 | 0.048 | 0.040 | 0.538 |
|  | 500 | 0.200 | 0.027 | 0.022 | 0.018 | 0.938 |
|  | 1000 | 0.199 | 0.019 | 0.016 | 0.013 | 0.987 |
|  | 50 | 0.204 | 0.091 | 0.067 | 0.056 | 0.321 |
| $r = 0.1$ | 100 | 0.205 | 0.068 | 0.049 | 0.040 | 0.501 |
|  | 500 | 0.201 | 0.031 | 0.022 | 0.018 | 0.893 |
|  | 1000 | 0.201 | 0.022 | 0.016 | 0.013 | 0.959 |

Table 6.3 gives the results for $T = 50, 100, 500, 1000$, $r = 1$ or 0.1, each based on 1000 simulations of the $T$TRWS. The mean of the naïvie SE (square root of variance) estimates is given, which clearly underestimates the true SE($\hat{\mu}$). Notice that in the current setting, as the length of walk $T$ increases, the absolute bias of naïve SE estimation is reduced but not the relative bias. Without applying any adaptive observation procedure, TRWS entails a loss of efficiency compared to standard SRS, as can be seen from SE-IID based on an IID sample.

In the population graph $G$, the case nodes have larger degrees than the noncase nodes, such that the stationary probability $\pi_i$ depends on $y_i$ and TRWS is informative in this sense. Nevertheless, informative sampling as such is not an issue for design-based estimation of graph parameters.

The consistency of $\hat{\mu}$ under TRWS is already evident at $T = 50$, even without removing the initial burn-in states. The last column shows the average of the traverse $\psi$ over the simulations. In the current setting, a TRW of length $T = 50$ is expected to visit only about a third of the 100 nodes in the population graph, whereas even a walk of length $T = 1000$ can not always reach all the nodes.

How quickly TRW traverses the population graph depends largely on the isolated nodes that can only be visited by random jumps, the probabilities of which are reduced given small $r$. The TRW traverses somewhat more slowly given $r = 0.1$ than $r = 1$, as is the convergence of $\hat{\mu}$, which can be seen by comparing the corresponding $\mathrm{SE}(\hat{\mu})$ given $r = 0.1$ or $r = 1$.

### 6.4.3    Estimation of a 3rd-order graph parameter

Let $\mu = \theta/\theta'$, where $\theta$ is the total number of triangles made up of case nodes, and $\theta'$ is that of the other triangles with at least one noncase node. The larger the value of $\mu$, the higher is the transitivity among cases compared to the overall transitivity in the graph. We have $\mu = 4.667$ in this population graph.

Figure 6.3    Illustration for triangle motifs under TRW

Let $\Omega$ contain all the triangles in $G$. Given $(X_t, X_{t+1}) = (i, j)$ of two adjacent nodes under TRW, one observes all the triangles involving $i$ and $j$. As illustrated in Figure 6.3, both the triangles $[\star ij]$ and $[\diamond ij]$ are observed from $(X_t, X_{t+1}) = (i, j)$. For (6.13), we have $\delta_M = 1$ iff $M = (i, j)$ and $I_\kappa(M) = 1$ if $\kappa$ is either of these two triangles.

Given $(X_t, X_{t+1}) = (i, j)$ as the AS3, the ES3s are the six possible adjacent moves along the triangle. Under TRW (6.6) and

(6.7), the S3P is

$$\pi_i p_{ij} \propto 1 + \frac{r}{N}$$

such that (6.14) and (6.15) yield the same weight $w_{M\kappa} \equiv 1/6$.

Table 6.4  Estimation of $\mu = \theta/\theta' = 4.667$ under $T$TRWS

| $T$ | Mean($\hat{\mu}$) | SD($\hat{\mu}$) | $\psi$ |
|---|---|---|---|
| | $r = 0.1$ | | |
| 100 | 6.119 (0.142) | 4.498 | 0.498 |
| 500 | 4.737 (0.028) | 0.893 | 0.893 |
| 1000 | 4.669 (0.019) | 0.593 | 0.958 |
| $T$ | Mean($\hat{\mu}$) | SD($\hat{\mu}$) | $\psi$ |
| | $r = 6$ | | |
| 100 | 6.362 (0.160) | 5.069 | 0.606 |
| 500 | 4.805 (0.034) | 1.075 | 0.983 |
| 1000 | 4.704 (0.022) | 0.702 | 0.999 |

Table 6.4 gives the results for $T = 100, 500, 1000$ and $r = 0.1, 6$, each based on 1000 simulations of $T$TRWS. The initial probabilities are $p_{0,i} = \pi_i$ directly. There is a small probability that no triangle is observed when $T = 50$, in which case $\hat{\mu}$ cannot be computed. Any choice of $T < 100$ is thus omitted here.

Since the average degree is about 6 in the population graph here, setting $r = 6$ in (6.6) makes a random jump on average at least as probable as an adjacent move at each time step. This raises the traverse of the walk, e.g. TRW of length $T = 1000$ can almost be expected to cover the whole population graph.

The convergence of $\hat{\mu}$ seems not greatly affected by $r$, although intuitively a large value of $r$ is unlikely to be efficient, as the order of motif increases. Given either $r$, convergence is more or less the case for $T = 1000$, where each value in the parentheses is the estimated simulation error of Mean($\hat{\mu}$). Clearly, the walk needs to be longer for estimating this 3rd-order graph parameter than for the population case prevalence. This is not surprising, because not every two successive states $(X_t, X_{t+1})$ correspond to an adjacent move, nor does one necessarily observe any triangle based on every adjacent move. In contrast, every $X_t$ contributes to the estimation of the population case prevalence.

## BIBLIOGRAPHIC NOTES

Masuda et al. (2017) give a comprehensive review of random walks and diffusion on networks. Boyd et al. (2004) provide results of the convergence rate for pure random walks in a graph. No general results seem to be available for targeted walks, where the stationary probability is not exclusively determined by the degree of the node in undirected graphs and random jumps are allowed in addition.

Salganik and Heckathorn (2004) propose RDS for hard-to-catch populations, inspired by random walk in graphs. It is unrealistic that respondent-driven observation can satisfy the requirement of the random-walk OP. In applications, therefore, the estimation and inference need to rely on model assumptions instead.

Until recently, the sampling theory of random walk in graphs has only dealt with the 1st-order graph parameters, such as the population case prevalence. Thompson (2006a) uses the generalised ratio estimator under random walk with jumps and targeted MH walks. The practical algorithm of TRW (6.6) and (6.7) is given by Avrachenkov et al. (2010).

Thompson (2006b) develops adaptive web sampling targeted at 1st-order graph parameters, which is a hybrid of breadth-first snowball sampling and depth-first walk sampling.

The strategy BIGS-IWE can be adapted to TRWS sequence-by-sequence. For the 3rd-order graph parameter illustrated above, the two basic choices of incidence weight coincide. But this is not the general case. For instance, let $[\{i, j, g, h\}]$ be a 4-cycle, which can be observed based on AS3 consisting of two successive adjacent moves, say, $(X_t, X_{t+1}, X_{t+2}) = (i, j, g)$. The S3P is

$$\pi_i p_{ij} p_{jg} = \frac{1}{d_j + r} \left(1 + \frac{r}{N}\right)^2$$

which varies with the degree of the second node in the sequence.

# Bibliography

[1] Avrachenkov, K., Ribeiro, B. and Towsley, D. (2010). Improving Random Walk Estimation Accuracy with Uniform Restarts. *Research report*, RR-7394, INRIA. inria-00520350

[2] Becker, E.F. (1991). A terrestrial furbearer estimator based on probability sampling. *The Journal of Wildlife Management*, 55:730–737.

[3] Birnbaum, Z.W. and Sirken, M.G. (1965). *Design of Sample Surveys to Estimate the Prevalence of IRareDiseases: Three Unbiased Estimates.* Vital and Health Statistics, Ser. 2, No.11. Washington:Government Printing Office.

[4] Blackwell, D. (1947). Conditional expectation and unbiased sequential estimation. *The Annals of Mathematical Statistics*, 18:105–110.

[5] Boyd, S., Diaconis, P. and Xiao, L. (2004). Fastest mixing Markov chain on a Graph. *SIAM Review*, 46:667–689.

[6] Brin, S. and Page, L. (1998). The anatomy of a large-scale hypertextual web search engine. *Computer Networks and ISDN Systems*, 30:107–117.

[7] Chung, F.R.K. (1997). *Spectral Graph Theory.* Providence, RI: American Mathematical Society.

[8] Cochran, W.G. (1977). *Sampling Techniques (3rd ed.).* New York: Wiley.

[9] Dryver, A.L. and Thompson, S.K. (2005). Improved unbiased estimators in adaptive cluster sampling. *Journal of the Royal Statistical Society, Series B*, 67:157–166.

[10] Frank, O. (1971). *Statistical inference in graphs.* Stockholm: Försvarets forskningsanstalt.

[11] Frank, O. (1977a). Estimation of graph totals. *Scandinavian Journal of Statistics*, 4:81–89.

[12] Frank, O. (1977b). A note on Bernoulli sampling in graphs and Horvitz-Thompson estimation. *Scandinavian Journal of Statistics*, 4:178–180.

[13] Frank, O. (1977c). Survey sampling in graphs. *Journal of Statistical Planning and Inference*, 1:235–264.

[14] Frank, O. (1978). Estimation of the number of connected components in a graph by using a sampled subgraph. *Scandinavian Journal of Statistics*, 5:177–188.

[15] Frank, O. (1979). Sampling and estimation in large social networks. *Social networks*, 1:91–101.

[16] Frank, O. (1980a). Estimation of the number of vertices of different degrees in a graph. *Journal of Statistical Planning and Inference*, 4:45–50.

[17] Frank, O. (1980b). Sampling and inference in a population graph. *International Statistical Review*, 48:33–41.

[18] Frank, O. (1981). A survey of statistical methods for graph analysis. *Sociological methodology*, 12:110–155.

[19] Frank, O. (2011). Survey sampling in networks. *The SAGE Handbook of Social Network Analysis*, pp. 389–403.

[20] Frank O. and Snijders T. (1994). Estimating the size of hidden populations using snowball sampling. *Journal of Official Statistics*, 10:53–53.

[21] Godambe, V.P. and Joshi, V.M. (1965). Admissibility and Bayes Estimation in Sampling Finite Populations. *The Annals of Mathematical Statistics*, 36:1707–1722.

[22] Goldenberg, A., Zheng, A.X., Fienberg, S.E. and Airoldi, E.M. (2010). A Survey of Statistical Network Models. *Foundations and Trends® in Machine Learning*, 2:129–233.

[23] Goodman, L.A. (1961). Snowball sampling. *Annals of Mathematical Statistics*, 32:148–170.

[24] Horvitz, D.G. and Thompson, D.J. (1952). A generalization of sampling without replacement from a finite universe. *Journal of the American Statistical Association*, 47:663–685.

[25] Hu, P. and Lau, W.C. (2013). A Survey and Taxonomy of Graph Sampling. https://arxiv.org/pdf/1308.5865.pdf

[26] Kaiser, L. (1983). Unbiased estimation in line-intercept sampling. *Biometrics*, 39:965–976.

[27] Klovdahl, A.S. (1989). Urban social networks: Some methodological problems and possibilities. In M. Kochen (ed.) *The Small World*. Norwood, NJ: Ablex Publishing, pp. 176–210.

[28] Lavallee, P. (2007). *Indirect Sampling*. Springer.

[29] Leskovec, J. and Faloutsos, C. (2006). Sampling from Large Graphs. *In Proceedings of ACM SIGKDD*.

[30] Masuda, N., Porter, M.A. and Lambiotte, R. (2017) Random walks and diffusion on networks. *Physics Reports*, 716–717:1–58. http://dx.doi.org/10.1016/j.physrep.2017.07.007

[31] Newman, M.E.J. (2010). *Networks: An Introduction*. Oxford University Press.

[32] Neyman, J. (1934). On the two different aspects of the representative method: The method of stratified sampling and the method of purposive selection. *Journal of the Royal Statistical Society*, 97:558–606.

[33] Patone, M. (2020) *Topics of Statistical Analysis with Social Media Data*. Unpublished PhD Thesis.

[34] Patone, M. and Zhang, L.-C. (2020) Incidence weighting estimation under bipartite incidence graph sampling. arXiv: 2004.04257v1

[35] Rao, C.R. (1945). Information and accuracy attainable in the estimation of statistical parameters. *Bulletin of Calcutta Mathematical Society*, 37:81–91.

[36] Salganik, M.J. and Heckathorn, D.D. (2004). Sampling and estimation in hidden populations using respondent-driven sampling. *Sociological Methodology*, 34:193–239.

[37] Sirken, M.G. (1970). Household surveys with multiplicity. *Journal of the American Statistical Association*, 65:257–266.

[38] Sirken, M.G. (2004). Network sample survey of rare and elusive populations: a historical review. In *Preceedings of Statistics Canada Symposium: Innovative Methods for Surveying Difficult-to Reach Population.*

[39] Sirken, M.G. (2005). *Network Sampling.* In *Encyclopedia of Biostatistics*, John Wiley & Sons, Ltd. DOI: 10.1002/0470011815.b2a16043

[40] Sirken, M.G. and Levy, P.S. (1974). Multiplicity estimation of proportion based on ratios of random variable. *Journal of the American Statistical Association*, 69:68–73.

[41] Snijders, T.A.B. (1992). Estimation on the basis of snowball samples: How to weight. *Bulletin de Methodologie Sociologique*, 36:59–70.

[42] Thompson, S.K. (1990). Adaptive cluster sampling. *Journal of the American Statistical Association*, 85:1050–1059.

[43] Thompson, S.K. (1991). Adaptive cluster sampling: Designs with primary and secondary units. *Biometics*, 47:1103–1115.

[44] Thompson, S.K. (2006a). Targeted random walk designs. *Survey Methodology*, 32:11–24.

[45] Thompson, S.K. (2006b). Adaptive Web Sampling. *Biometrics*, 62:1224–1234.

[46] Thompson, S.K. (2012). *Sampling.* John Wiley & Sons, Inc.

[47] Vincent, K. and Thompson, S.K. (2017). Estimating population size with link-tracing sampling. *Journal of the American Statistical Association*, 112:1286–1295.

[48] Zhang, L.-C. (2021). Graph sampling: An introduction. *The Survey Statistician*, 83:27–37.

[49] Zhang, L.-C. (2020). Sampling designs for epidemic prevalence estimation. https://arxiv.org/abs/2011.08669

[50] Zhang, L.-C. and Oguz-Alper, M. (2020). Bipartite incidence graph sampling. https://arxiv.org/abs/2003.09467

[51] Zhang, L.-C. and Patone, M. (2017). Graph sampling. *Metron*, 75:277.

# Index

Note: Page numbers in *italics* indicate figures and **bold** indicate tables in the text.